DATA MINING AND DATA WAREHOUSING

By

GUNJAN GOSWAMI
Assistant Professor, Department of CSE,
JSS Academy of Technical Education,
Noida, U.P.

S.K. KATARIA & SONS®
Publisher of Engineering Books
4885/109, Prakash Mahal, Dr. Subhash Bhargav Lane,
Opposite Delhi Medical Association, Daryaganj, New Delhi–110002 (INDIA)
Phone: +91-11-43551243; *Mobile:* +91-9871775858, 7678425209
e-mail: katariabook@gmail.com, katariabooks@yahoo.com
Website: www.skkatariaandsons.com
(e-books, subscriptions and bulk buying also available on our website)

Published By:

S.K. Kataria & Sons®
4885/109, Prakash Mahal, Dr. Subhash Bhargav Lane,
Opposite Delhi Medical Association, Daryaganj, New Delhi-110002 (INDIA)
Phone: +91-11-43551243; *Mobile:* +91-9871775858, 7678425209
e-mail: katariabook@gmail.com, katariabooks@yahoo.com
Website: www.skkatariaandsons.com

First Edition: 2012; Second Edition: 2013; Reprint: 2026

Price: ₹ 395/-

Laser Typesetting by:
AISA Enterprises, Delhi-110092

Printed at :
Fine Offset Printers

Preface

Many areas of science, business, and other environments deal with a vast amount of data, which needs to be turned into something meaningful, knowledge. The data we have is often vast, and noisy, meaning that it's imprecise and the data structure is complex. This is where a purely statistical technique would not succeed, so data mining is a solution.

From www if we want to search on any topic, then we get lots of links for that topic. But out of this the "useful" information is less. Thus as the data is going on increasing, the useful information is going on decreasing. Thus something is needed that will automatically: analyze the data, summarize the data, classifies the data, cluster the data, remove all kinds of anomalies etc. To automate these, data mining is required. Data Mining (DM) techniques are useful in biological treatment, fraud detection, pattern recognition, mining web etc.

Data Mining has to mine the data but form where? Answer is Data Warehouse. A data warehouse subject-oriented, integrated, time-variant, and nonvolatile, collection of data that support of the management's decision making process. Thus Data Warehouse is developed to meet the growing demand for information analysis that could not be met by traditional operational systems.

Organization of the book:

The purpose of this book is to introduce various Data Warehousing and Data Mining concepts and algorithms. The book is organized into 11 chapters.

Chapter 1: Introduction to data warehousing: This chapter provides the basics of data warehousing like what is Data Warehouse. Why this is needed, Advantages, disadvantages, applications etc.

Chapter 2: Data Warehouse: features and architecture: This chapter defines the basic features and architecture of DW system. It also discusses data granularity, concept of Meta data etc. further in this chapter different types of architectures along with their advantages and disadvantages are discussed.

Chapter 3: Dimensional Modeling: In this chapter different data warehouse schemas are discussed with their pros and cons. Also this chapter covers the characteristics of the fact tables, dimension table etc.

Chapter 4: OLAP strategies for large data warehouse: In this chapter various aspects of Online Analytical Processing (OLAP) technology are discussed. ROLAP, MOLAP, difference between them, OLAP servers are also covered in this chapter.

Chapter 5: Data warehouse implementation: This chapter deals with data warehouse implementation issues. Since DW contains large volumes of data and it is in demand that the queries must be answered in seconds. So in this chapter various techniques are studied for computing the data cube efficiently. Also Backend tools and utilities , testing of data warehousing is also discussed in this chapter.

Chapter 6: Introduction to data mining: This chapter introduces the concept of data mining technology, tools and applications. Data Mining tasks, KDD vs. DM, the need of DM, issues etc. have been covered in this chapter.

Chapter 7: Data Mining Query Language (DMQL): This chapter covers the concept of Data Mining Query Language. The syntax that is how to write the queries is discussed in this chapter.

Chapter 8: Classification: This chapter deals with classification and prediction. Classification is done with the help of various algorithms, techniques etc. Various techniques like Back propagation algorithm, Genetic algorithm etc. are covered in this chapter.

Chapter 9: Association rules: this chapter explain the association rule mining, association rules to mine single dimensional from transactional databases, mine the multilevel association rules from transactional databases and so on.

Chapter 10: Cluster Analysis: this chapter deals with grouping the items having same attributes in one group and dissimilar items are excluded from this. Various categories of clustering methods like partitioning methods, hierarchical clustering, density based methods and outlier analysis are covered in this chapter.

Chapter 11: Mining complex types of data: this chapter discusses the different techniques of data mining like text mining, web mining, spatial databases etc.

AUTHOR

Acknowledgement

The book itself is an acknowledgement to the intensity, drive and technical competence of many individuals, who have contributed to it.

I am thankful to **Almighty God**, without whose grace nothing is possible for any one. I am very thankful to my friends and students who have inspired me to write this book.

I am grateful to **my family members** for their constant motivation and guidance that has helped me to complete this book. I would also express my gratefulness for their patience and understanding during the time I devoted to writing the book.

I would like to take this opportunity to express my thanks to **Mr. Sanjeev Kumar Kataria** of M/s S.K. Kataria and Sons for publishing this book in such a short time. I am also thankful to **Mr. Sachin** for his assistance.

Gunjan Goswami
Author

Dedicated to

Almighty God

Contents

1. INTRODUCTION TO DATA WAREHOUSING .... 1—17

1.1. Introduction 1
1.2. Data Warehouse 2
1.3. History of Data Warehousing 2
1.4. Data Warehouse Definitions 3
  1.4.1. What can a Data Warehouse do? 3
  1.4.2. What can a Data Warehouse not do? 4
1.5. Characteristics of a Data Warehouse 4
1.6. Need of Data Warehouse 5
1.7. Benefits of a Data Warehouse 6
1.8. DBMS Vs. Data warehouse 6
1.9. Components, Architecture, Design and Construction of Data Warehouse 9
1.10. Data Warehouse Architecture 10
  1.10.1. Data Warehouse Architecture (Basic) 10
  1.10.2. Data Warehouse Architecture (With a Staging Area) 11
  1.10.3. Data Warehouse Architecture (With a Staging Area and Data Marts) 12
1.11. ETL Process 12
1.12. Advantages and Disadvantages of Data Warehouse 15
  Summary 16
  Review Questions 17

2. DATA WAREHOUSE: FEATURES AND ARCHITECTURE . 18—36

2.1. Features of Data Warehouse 18
  2.1.1. Subject-oriented 19
  2.1.2. Integrated 19
  2.1.3. Time-variant 19
  2.1.4. Nonvolatile 20
  2.1.5. Data Granularity 20
2.2. Data Granularity 20
  2.2.1. Advantages and Disadvantages of Data Granularity 21
2.3. MetaData 22
  2.3.1. What is Metadata? 22
  2.3.2. Need of Meta Data 23
  2.3.3. Characteristics of Metadata Architecture . 24
  2.3.4. Metadata Lifecycle 25
  2.3.5. Metadata Repository 25
2.4. Architecture of Data Warehouse 26
  2.4.1. Design of Data Warehouse 26
  2.4.2. Two Tier Data Warehouse Architecture—Independent Data Marts ... 27
  2.4.3. Three Tier Architecture 27
2.5. Data Warehouse Models 29
  2.5.1. Enterprise Warehouse 29
  2.5.2. Virtual Warehouse 30
2.6. Data Marts 30
  2.6.1. Similarities and Dissimilarities Between DW and Data Marts 31
  2.6.2. Dependent and Independent Data Marts . 31
  2.6.3. The Steps in Implementing a Data Mart .. 33
  Summary 34
  Review Questions 36

3. DIMENSIONAL MODELING .... 37—51

3.1. Introduction 37
3.2. Dimensional Model vs ER Model 38
  3.2.1. Need for a DM 38
3.3. Data Warehouse Schema Architecture 39
  3.3.1. Star Schema 39
  3.3.2. Snowflake Schema 40
  3.3.3. Fact Constellation Schema/ Glaxy Schema 42
3.4. A Data Mining Query Language, DMQL: Language Primitives 43
3.5. Aggregate Tables 44
3.6. Data Cubes 45
  3.6.1. Representation of Data Cubes 46
  3.6.2. Operations on Data Cubes 47
3.7. Query Processing 47
  3.7.1. Indexing 47
3.8. Strengths of Dimensional Model 49
3.9. ER vs DM 50
  Summary 51
  Review Questions 51

4. OLAP STRATEGIES FOR LARGE DATA WAREHOUSE PROJECTS .... 52—63

4.1. Introduction 52
4.2. The Need For Multidimensional Information 53
4.3. OLAP and OLTP 54
4.3.1. OLTP vs. OLAP 54
4.3.2. Differences Between OLTP and OLAP System 55
4.3.3. Twelve Rules for OLAP by CODD 55
4.4. OLAP operations in Multidimensional Data Model 56
4.5. OLAP Server/Models 57
4.6. MOLAP—Multidimensional OLAP 57
4.7. ROLAP—Relational OLAP 58
4.8. Difference between ROLAP and MOLAP 59
4.9. HOLAP—Hybrid OLAP 60
4.10. DOLAP—Desktop OLAP 61
4.11. OLAP Applications 61
Summary 61
Review Questions 63

5. DATA WAREHOUSE IMPLEMENTATION .... 64—80

5.1. Introduction 64
5.1.1. Efficient Data Cube Computation 64
5.1.2. Lattice of Cuboids 64
5.2. Materialization 66
5.3. Cube Computation 67
5.3.1. ROLAP-Based Cubing Algorithms 67
5.3.2. MOLAP Base Cubing Algorithm 67
5.3.3. Iceberg Cube 68
5.3.4. Indexing OLAP Data 69
5.3.5. Data Warehousing Tools 71
5.4. Complex Aggregation at Multiple Granularities: Multifeature Cube 72
5.4.1. Other Methods 73
5.5. Testing and Tuning the Data Warehouse 73
5.5.1. Challenges of Data Warehouse Testing: 73
5.5.2. Business Users Test Business Intelligence 74
5.5.3. Tuning The Data Warehouse 75
5.5.4. Tuning Strategies 75
5.6. Query Manager 77
5.7. Case Study: The Data Warehouse Manager 78
5.7.1. The Data Warehouse Center 78
5.8. Query Management Facility 79
5.9. Future of Data Warehousing 79
Summary 79
Review Questions 80

6. INTRODUCTION TO DATA MINING .... 81—102

6.1. Overview 81
6.2. History of Data Mining 82
6.3. Defining Data Mining 83
6.4. Basic Data Mining Tasks 84
6.4.1. Prediction 84
6.4.2. Description 84
6.5. KDD Vs. Data Mining 85
6.6. Data Mining Application 88
6.7. The Scope of Data Mining 90
6.7.1. Automated Prediction of Trends and Behaviours 90
6.7.2. Automated Discovery of Previously Unknown Patterns 90
6.8. Data Mining Techniques 91
6.9. Why Use Data Mining? 92
6.9.1. How Data Mining Works 92
6.10. Data Mining Tools Comparison 92
6.11. Advantages and Disadvantages of Data Mining 96
6.11.1. Advantages of Data Mining 96
6.11.2. Disadvantages of Data Mining 97
6.12. Applications of Data Mining 97
6.13. KDD-Data Mining Issues and Challenges 99
Summary 101
Review Questions 102

7. DATA MINING QUERY LANGUAGE .... 103—114

7.1. Introduction 103
7.2. Importance of Data Mining Query Language 104
7.3. Syntax for Task-Relevant Data Specifications 104
7.4. DMQL: Syntax For Specification of Task-Relevant Data 106
7.5. Syntax for Specifying the Kind of Knowledge to be Mined 107
7.6. Syntax for Concept Hierarchy Specification 108
7.7. Syntax for Interestingness Measure Specification 108
7.8. Syntax for Pattern Presentation and Visualization Specification 109
7.9. Architecture of DM Systems 109
Summary 110
Review Questions 113

8. CLASSIFICATION .... 115—156

8.1. Introduction 115
8.2. Classification and Issues 115
8.2.1. Issues 118
8.3. Neural Networks 118
8.3.1. Where to Use Artificial Neural Networks 119
8.3.2. Neural Networks for Clustering 119
8.3.3. Neural Networks for Feature Extraction 119
8.3.4. How it Works 119
8.3.5. Models of Artificial Neural Networks ... 123
8.3.6. Neural Network Learning 123
8.4. Distance Based Algorithms 128
8.5. Case Based Reasoning (CBR) 129
8.6. Statistical Based Algorithm (Bayesian Classification) 129
8.6.1. Bayesian Classification 129
8.6.2. Bayes Theorem 129
8.7. Naive Bayes Classifier 130
8.7.1. The Naive Bayes Probabilistic Model ... 130
8.7.2. Bayes' Theorem Derived via Conditional Probabilities 132
8.7.3. Certainty Factor 135
8.8. Bayesian Network 135
8.9. Decision Tree Based Algorithm (ID3, C4.5, CART) 136
8.9.1. Advantages and Disadvantages of Decision Trees 137
8.9.2. Issues in Decision Trees Algorithms 138
8.9.3. Decision Tree Based Algorithms 139
8.10. Genetic Algorithm 142
8.10.1. Introduction to Genetic Algorithms (GA) 142
8.10.2. Working Principle of GA 142
8.10.3. Genetic Algorithms: Selection Operators 143
8.10.4. Genetic Algorithm: Crossover and Mutation Operators 144
8.10.5. Advantages of Limitations of GA 145
8.11. Rough Sets 145
8.11.1. A Rough Set Based KDD Process 146
8.12. Fuzzy Sets Approach 147
8.12.1. Advantages and Disadvantages of Fuzzy Logic 148
8.13. Support Victor Machine 148
8.13.1. Properties of SVM 153
Summary 154
Review Questions 156

9. ASSOCIATION RULES .... 157—173

9.1. Introduction 157
9.2. Concept of Frequent Itemsets, Closed Itemsets and Association Rules 158
9.3. Rule Strength Measures 158
9.3.1. Support 159
9.4. Classification of Association Rules 160
9.4.1. Based on the Completeness of Patterns to be Mined 160
9.4.2. Based on the Levels of Abstraction Involved in the Rule Set 160
9.4.3. Based on the Number of Data Dimensions Involved in the Rule 160
9.4.4. Based on the Types of Values Handled in the Rule 160
9.4.5. Based on the Kinds of Rules to be Mined 161
9.4.6. Based on the Kinds of Patterns to be Mined 161
9.5. Basic Algorithms For Mining Single Dimensional Boolean Association Rules From Transactional Databases 161
9.5.1. Apriori Algorithm 161
9.5.2. FP-Growth Algorithm 166
9.6. Iceberg Queries 167
9.7. Different Approaches For Mining Multilevel Association Rule 168
9.8. Mining Multi Dimensional Association Rules from Relational Databases and Data Warehouse 169
9.9. Basic Approaches for Mining Multidimensional Association Rules 169
Summary 172
Review Questions 173

10. CLUSTERING .... 174—187

10.1. Introduction 174
10.2. Applications of Clustering 175
10.3. Requirements of Clustering in Data Mining 176
10.4. Categories of clustering methods 177
10.4.1. Partitioning Algorithms 177
10.4.2. Hierarchical Methods 180
10.4.3. Density Based Clustering Methods 182
10.4.4. Model Based Clustering Methods 185
10.5. Outlier Analysis 185
Summary 186
Review Questions 187

11. MINING COMPLEX TYPES OF DATA .... 188—214

11.1. Introduction 188
11.2. Spatial Database Mining 188
11.2.1. Characteristics of Spatial Data Mining 189
11.2.2. Uses of Spatial DM 190
11.2.3. Techniques in Spatial DM 190
11.2.4. Spatial Clustering Methods 191
11.2.5. Spatial Classification and Spatial Trend Analysis 191
11.3. Web Mining 192
11.3.1. Web Content Mining 193
11.3.2. Web Usage Mining 194
11.3.3. Web Structure Mining 195
11.4. Text Mining 196
11.4.1. Data Mining Vs. Text Mining 197
11.4.2. Information Retrieval 197
11.4.3. Different Text Mining Approaches 199
11.4.4. Text Mining Application 200
11.5. Multimedia Databases Mining 201
11.5.1. Multimedia Indexing and Retrieval System 202
11.6. Mining Time Series and Sequence Data 204
11.6.1. Noisy and Missing Data 204
11.7. Trends in Data Mining 206
11.8. Mining Time Series Data 206
11.8.1. Two Aspects of Time Series Analysis 209
11.9. Complex Data Objects Mining 209
11.9.1. Generalization of Object Identifiers and Class/Subclass Hierarchies 210
11.10. Sequence Pattern Mining 211
11.10.1. Parameters for Sequential Pattern Mining 211
11.10.2. Approaches for Sequential Pattern Mining 211
- *Summary* 212
- *Review Questions* 213

QUESTION BANK .... 215—247

TEST YOURSELF .... 248—251

MODEL TEST PAPERS .... 252—256

EXAMINATION PAPERS .... 257—258

INDEX .... 259—262

1

Chapter

Introduction to Data Warehousing

INSIDE THIS CHAPTER

1.1. Introduction **1.2.** Data Warehouse **1.3.** History of Data Warehousing **1.4.** Data Warehouse Definitions **1.5.** Characteristics of a Data Warehouse **1.6.** Need of Data Warehouse **1.7.** Benefits of a Data Warehouse **1.8.** DBMS Vs. Data Warehouse **1.9.** Components, Architecture, Design and Construction of Data Warehouse **1.10.** Data Warehouse Architecture **1.11.** ETL Process **1.12.** Advantages and Disadvantages of Data Warehouse

This chapter aims to provide an overview of fundamental concepts of data warehousing. The definition, its evolution, characteristics, advantages, disadvantages are described in this chapter.

1.1. INTRODUCTION

A Data Warehouse is a repository of integrated information, available for queries and analysis. Data and information are extracted from heterogeneous sources as they are generated. This makes it much easier and more efficient to run queries over data that originally came from different sources. It contains intelligent data collections which are modelled to support the reporting and analysis needs of DSS users.

A data warehouse is a relational database that is designed for query and analysis rather than for transaction processing. It usually contains historical data derived from transaction data, but it can include data from other sources. It separates analysis workload from transaction workload and enables an organization to consolidate data from several sources. A data warehouse environment includes an extraction, transportation, transformation, and loading (ETL) solution, an online analytical processing (OLAP) engine, client analysis tools, and other applications that manage the process of gathering data and delivering it to business users. It is expected to present the right information in the right place at the right time with the right cost in order to support the right decision. The practice proved that the traditional online Transaction Processing (OLTP) systems are not appropriate for decision support and the high speed networks can not by themselves solve the information accessibility problem.

1.2. DATA WAREHOUSE

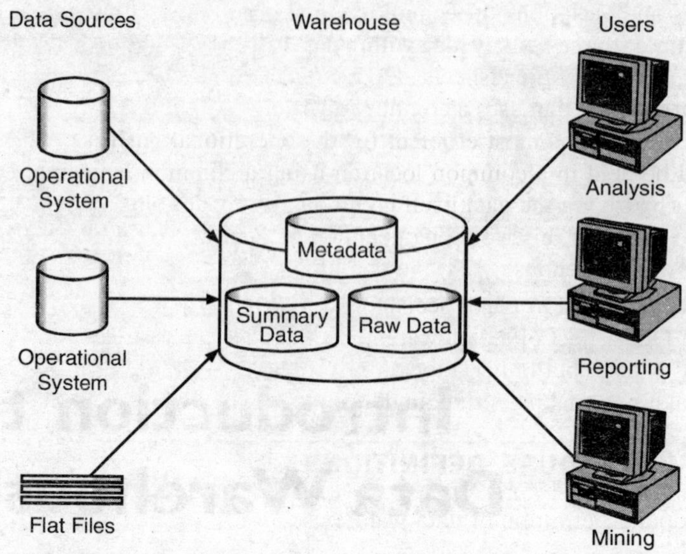

Fig. 1.1. Data Warehouse.

Data warehousing has become an important strategy to integrate heterogeneous information sources in organizations, and to enable Online Analytic Processing. This movement is a consequence of the observation by W. Inmon and E.F. Codd in the early 1990's that operational-level on-line transaction processing (OLTP) and decision support applications (OLAP) cannot coexist efficiently in the same database environment, for two main reasons, both having to do with trade-offs in data quality.

A data warehouse is a collection of data gathered and organized so that it can easily by analyzed, extracted, synthesized, and otherwise be used for the purposes of further understanding the data. It may be contrasted with data that is gathered to meet immediate business objectives such as order and payment transactions, although this data would also usually become part of a data warehouse. Data warehouses can be subdivided into data marts. Data marts store subsets of data from a warehouse

A data warehouse brings together data from multiple transactional systems and enables users to access and analyze the information at various levels, from departmental to divisional, throughout the enterprise. A data warehouse is a record of an enterprise's past transactional and operational activities, stored in a database. The database design favours data analysis and reporting in order to gain strategic insight and to facilitate decision making. Data warehouses are not used for current, "live" data. A Data warehouse is a Relational Database that has assimilated data from various OLTP systems and other sources used for analytics.

1.3. HISTORY OF DATA WAREHOUSING

In the 1990's, as organizations of scale began to need more timely data about their business, they found that traditional information systems technology was simply too cumbersome to provide relevant data efficiently and quickly. Completing reporting requests could take days or weeks using antiquated reporting tools that were designed more or less to 'execute' the business rather than 'run' the business.

From this idea, the data warehouse was born as a place where relevant data could be held for completing strategic reports for management. The key here is the word 'strategic' as most executives were less concerned with the day to day operations than they were with a more overall look at the model and business functions.

As with all technology, over the course of the latter half of the 20th century, we saw increased numbers and types of databases. Many large businesses found themselves with data scattered across multiple platforms and variations of technology, making it almost impossible for any one individual to use data from multiple sources. A key idea within data warehousing is to take data from multiple platforms/ technologies (As varied as spreadsheets, DB2 databases, IDMS records, and VSAM files) and place them in a common location that uses a common querying tool. In this way operational databases could be held on whatever system was most efficient for the operational business, while the reporting / strategic information could be held in a common location using a common language. Data Warehouses take this even a step further by giving the data itself commonality by defining what each term means and keeping it standard. (An example of this would be gender which can be referred to in many ways, but should be standardized on a data warehouse with one common way of referring to each gender).

All of this was designed to make decision support more readily available and without affecting day to day operations. One aspect of a data warehouse that should be stressed is that it is NOT a location for ALL of a businesses data, but rather a location for data that is 'interesting'. Data that is interesting will assist decision makers in making strategic decisions relative to the organization's overall mission.

1.4. DATA WAREHOUSE DEFINITIONS

Two of the popular definition of data warehouse are :

1. A process of transforming data into information and making it available to users in a timely enough manner to make a difference. [*Forrester Research, April 1996*]
2. A single, complete and consistent store of data obtained from a variety of different sources made available to end users in a what they can understand and use in a business context. [*Barry Devlin*]

The collection of data and integration of data across sources is shown in the Fig. 1.2.

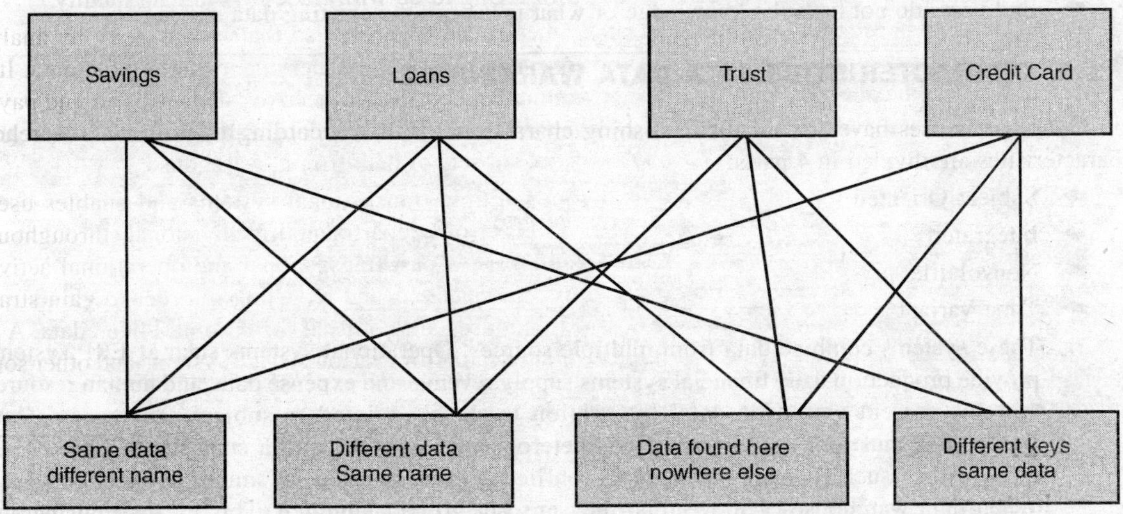

Fig. 1.2. Data Integration Across Sources.

1.4.1. What can a Data Warehouse do?

In this section the capabilities of data warehouse are described.

➡ **Get Answers Faster :** Data warehouse is helpful in getting the answers faster. Rapidly deliver information in context to decision makers and offer them an interactive, self-service environment for exploration and analysis

- **Make Decisions Faster :** Data warehouse provides decision makers with the right information and insights, where, when and how they need it
- **Optimize Performance :** Data warehouse easily measure and monitor financial and operational business performance, analyze results, predict outcomes and plan
- **Reduce Risk and Cost :** It means that appliances, pre-configured systems and warehouse accelerators reduce project risks, deployment time, resources and costs.

1.4.2. What can a Data Warehouse not do?

(1) **Can't create data itself :** Although data warehouse has the benefits, has the capabilities but still it has some limitations. It acts as an information repository that collects and reports data that already exists. But it cannot create its additional data on its own.

(2) **Cleaning of data is required :** If an organization has dirty data, the data warehouse will not be able to correct the results until and unless the data is first cleaned. Therefore in this context a data warehouse will only be able to identify where the problem exists but corrections will have to be made in the source systems that capture the data.

The main problem addressed by a data warehouse is that end-users have a difficult time producing ad-hoc or other specialized queries and reports. This is due to several factors:

- Most of the data is stored in ADABAS, which is difficult for end-users to access.
- The data stores were designed for transaction processing not for ad-hoc reporting.
- Obtaining the data or a report usually requires waiting for a programmer to either develop the report or provide a customized download program.
- All of the data may not be consistent as of the same point in time.
- There may not be enough copies of the data kept for historical reporting in the operational systems.
- End-users do not have the knowledge of what is kept in the existing data stores.

1.5. CHARACTERISTICS OF A DATA WAREHOUSE

Data warehouses have several distinguishing characteristics. But according to William Inmon the characteristics are divided in 4 parts:

- Subject Oriented
- Integrated
- Nonvolatile
- Time Variant

1. These systems combine data from multiple sources. Operational systems such as ERP systems provide production data, financial systems supply revenue and expense data, and human resource systems present employee data. Integration is closely related to subject orientation. Data warehouses must put data from various heterogeneous sources into a consistent format. They must resolve such problems as naming conflicts and inconsistencies among units of measure. When data warehouses achieve this, they are said to be integrated. This means that naming conflicts have to be resolved and problems like data being in different units of measure must be resolved.
2. The data copied into a data warehouse does not change (except to correct errors). This is logical because the purpose of a warehouse is to enable you to analyze what has occurred. The data warehouse is a historical record of the state of an organization. The frequent changes of the source OLTP systems are reflected in the data warehouse by adding new data, not by changing existing data. A data warehouse's focus on not to change the data is what meant by the term time variant.

3. Data warehouses are subject oriented, that is, they focus on measuring entities, such as sales, inventory, and quality. Using this warehouse, we can answer questions like "Who was our best customer for a particular item last year?" This ability to define a data warehouse by subject matter, sales in this case makes the data warehouse subject oriented. OLTP systems, by contrast, are function oriented and focus on operations such as order fulfillment.
4. In data warehouses, data from distinct function-oriented systems is integrated to provide a single view of an operational entity.
5. Data warehouses are designed for business users, not database programmers, so they are easy to understand and query.
6. Data warehouses are designed for speed. Users interact with data warehouses—running queries, analyzing data, formulating new queries based on information gleaned from earlier queries—so the systems must be responsive
7. Most business analysis requires analyzing trends. Because of this, analysts tend to need large amounts of data. This is very much in contrast to OLTP systems, where performance requirements demand that historical data be moved to an archive. In order to discover trends in business, analysts need large amounts of data. This is very much in contrast to online transaction processing (OLTP) systems, where performance requirements demand that historical data be moved to an archive. A data warehouse's focus on change over time is what is meant by the term time variant.

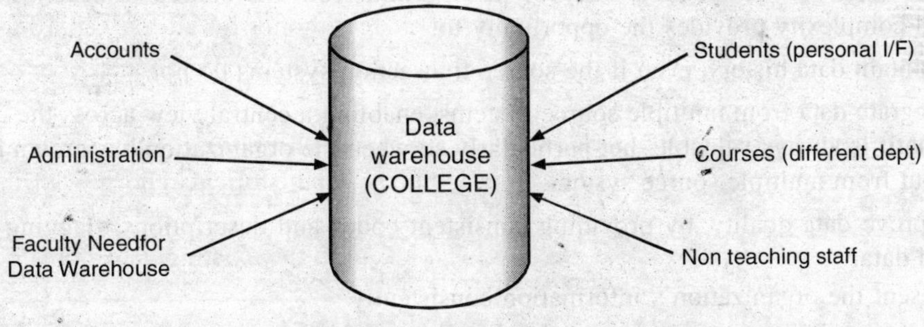

Fig. 1.3. Data Warehouse of a College.

1.6. NEED OF DATA WAREHOUSE

Data warehouse is needed for the following reasons :

1. **DW purpose for Data Consistency and Quality :** We know that in the organizations the information comes from different important information systems comes and each of these systems may carry the information in different formats and also may be having out of synchronous information. By bringing the data from these disparate sources at a common place, one can effectively undertake to bring the uniformity and consistency in data.
2. **Information Integration from Multiple Systems :** Data warehouse said to be the single point source for information. Data warehousing act as single point of source for all types of data stored and these information is in different format.
3. **High Response Time :** Production Databases are tuned to expected transaction load the Data-warehouse has to be ready for fairly unexpected loads and type of queries, which demands a high degree of flexibility and quick response time.
4. **High Response Time :** Normalized Data vs. Dimensional Modeling Production/Source system database are typically normalized to enable integrity and non-redundancy of data. This kind of design is fine for transactions, which involved few records at a time. However, for large analysis and mining queries, the response time in normalized databases will be slow given the joins that have to be created.

5. **Adaptive and Flexible Source of Information :** Data Warehouse objective is to provide an adaptive and flexible source of information. It is easier for users to define the production work and functionalities they want, but difficult to define the analysis they need. The analysis needs keep on changing and Data-Warehouse has the capabilities to adapt quickly to the changing requirements, but yes they can't create data itself. That's why they have the property of non volatility.
6. **Establish the Foundation for Decision Support :** In an organization the decision process involves analysis, data mining, forecasting, decision modeling etc. data warehousing provides a common point, for consistent, quality data with high response time that act as a core enabler for making fast and informed decisions.
7. **Keeping Analysis/Reporting and Production Separate :** Reporting and analysis work typically access data across the database tables, whereas production work typically accesses specific customer OR product OR channel record at a point of time. That's why it is important to have the Information generation work to be done from an offline platform. Purpose of Data Warehouse is to keep analysis/reporting (non-production use data) separate from production data.

1.7. BENEFITS OF A DATA WAREHOUSE

A data warehouse maintains a copy of information from the source transaction systems. This architectural complexity provides the opportunity to:

- Maintain data history, even if the source transaction systems do not.
- Integrate data from multiple source systems, enabling a central view across the enterprise. This benefit is always valuable, but particularly so when the organization has grown by merging the input from multiple source system.
- Improve data quality, by providing consistent codes and descriptions, flagging or even fixing bad data.
- Present the organization's information consistently.
- Provide a single common data model for all data of interest regardless of the data's source.
- Restructure the data so that it makes sense to the business users.
- Restructure the data so that it delivers excellent query performance, even for complex analytic queries, without impacting the operational systems.
- Add value to operational business applications, notably customer relationship management (CRM) systems

1.8. DBMS VS. DATA WAREHOUSE [MDU Dec. 2011]

Contrasting OLTP and Data Warehousing Environments, there is one major difference between the types of system. The difference is that data warehouses are not usually in **third normal form (3NF)**, a type of data normalization common in OLTP environments.

Data warehouses and OLTP systems have very different requirements. Here are some examples of differences between typical data warehouses and OLTP systems:

1. Workload

Data warehouses are designed to accommodate *ad hoc* queries. We might not know the workload of data warehouse in advance, so a data warehouse should be optimized to perform well for a wide variety of possible query operations.

OLTP systems support only predefined operations. Our applications might be specifically tuned or designed to support only these operations.

2. Data Modifications

A data warehouse is updated on a regular basis by the ETL process (run nightly or weekly) using bulk data modification techniques. The end users of a data warehouse do not directly update the data warehouse.

In OLTP systems, end users routinely issue individual data modification statements to the database. The OLTP database is always up to date, and reflects the current state of each business transaction.

3. Schema Design

Data warehouses often use denormalized or partially denormalized schemas (such as a star schema) to optimize query performance.

OLTP systems often use fully normalized schemas to optimize update/insert/delete performance, and to guarantee data consistency.

4. Typical Operations

A typical data warehouse query scans thousands or millions of rows. For example, "Find the total sales for all customers last month."

A typical OLTP operation accesses only a handful of records. For example, "Retrieve the current order for this customer."

5. Historical Data

Data warehouses usually store many months or years of data. This is to support historical analysis. OLTP systems usually store data from only a few weeks or months. The OLTP system stores only historical data as needed to successfully meet the requirements of the current transaction.

Contrasting OLTP and Data Warehousing Environments

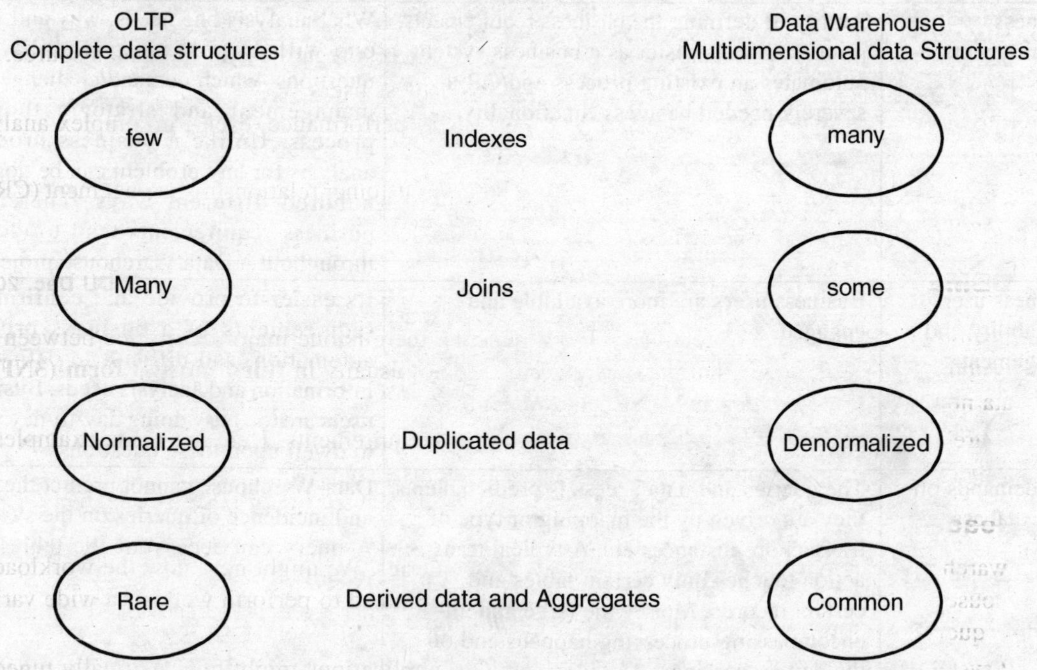

Fig. 1.4. OLTP and Data Warehouse.

Data Warehouse vs. OLTP Transaction Systems

Project Domain	Transaction/Production Business System	Data Warehouse System
Business Benefits	Tangible benefits in terms of functional capabilities, business processes that will be automated, number of headcounts and reduction etc. Typically, the process getting automated is being done manually, and there is enough visible pain at the ground level and customers.	There is a lesser proportion of initiatives where there is 'heaven will fall', if the project is not done. The benefits can be appreciated by fewer people and much fewer at a ground level.
Usage	A business system once implemented, drives the usage as it typically automates a business process.	Data Warehouse platform has a lesser compulsion for usage. Unless there are critical operational reports required.
Measure of Usage	One can specify the measure of usage for a business system in terms of processed unit, number of users.	While number of users and the number of queries does represent the level of usage, but it no means suggest that the usage is resulting in delivery of final outcome.
Skills and Expertise Requirements-Business	Business system requires the expertise on business process knowledge	More knowledge is required horizontally and vertically. One needs a much higher domain experience as well as cross-functional knowledge for an effective business role fulfillment in Data Warehouse project. The domain expertise also includes all three levels (strategic, managerial and operational).
Business Requirements	Ability of defining the business requirements, prioritization is easier as a business system automates an existing process and/OR a severely needed business functionality.	What analysis one needs, why and what one will do post its availability are questions, which demand/challenge the management and strategic thought process. Unlike a business process, analysis for any problem can be done in hundred different ways. Therefore, business requirements tend to change throughout a Data Warehouse project.
Business users availability and engagement	Business users are more available and engaged.	Its easier to provide and confirm the requirements of a business process automation, and difficult to define the information and analysis needs. Business users are too busy doing day-to-day work to dwell upon these questions.
The demands on the Database	The queries and data access is predictable as they are driven by the mapping of type of transaction, instances etc. A typical transaction touches only certain tables and certain records. Mostly the large and all-encompassing processing happens end of the day processing.	Data-Warehouse cannot predict the kind and incidence of queries on the system. A query can access all the tables and records.

Project Domain	Transaction/Production Business System	Data Warehouse System
Variety of front-end applications	A business system has a pre-defined back-end and front end applications accessing the back-end Database	A data-Warehouse could be having new front-end applications being added on the ongoing basis. This includes OLAP tools, Data mining applications, business performance management applications, online user query and reporting applications.
Expectations of flexibility to enhancements	A typical business system has an ever increasing list of enhancements, However, it is expected that the enhancements will take time and system will go through well-spaced out releases.	A Data-Warehouse is expected to provide granular enhancements for most cases. It has to have its design flexible enough to be able to incorporate new dimensions, measures and system sources without unsettling the foundations.

1.9. COMPONENTS, ARCHITECTURE, DESIGN AND CONSTRUCTION OF DATA WAREHOUSE

The components of the data warehouse are:

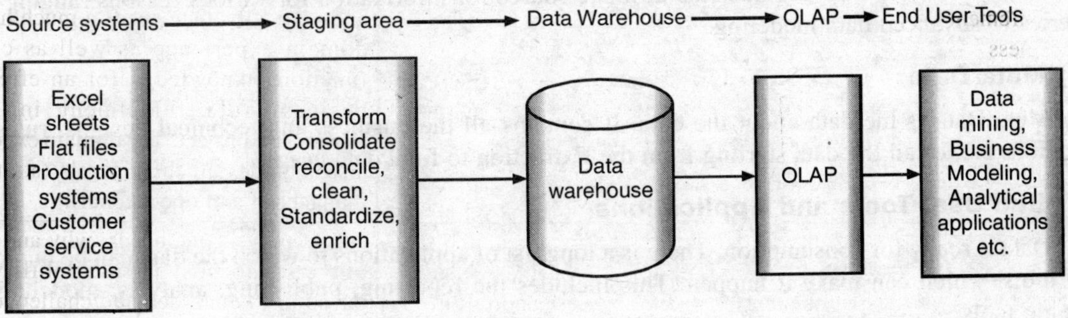

Fig. 1.5. Components of Data Warehouse.

(A) Source Systems and Databases

Source Systems is one of the major component of data warehouse. These systems are the raw data providers, from where the details are pulled out for making it suitable for Data Warehousing. The sources can be quite diverse like

- Production Databases like Oracle, SQL.
- Excel Sheets.
- Database of small time applications like in MS Access.
- ASCII/Data flat files.

(B) Data Staging 'Area'

In Data warehouse terms, a **data staging area** is an intermediate storage area between the sources of information and the data warehouse (DW) or Data mart (DM) as shown the fig above. It is usually of temporary nature, and its contents can be erased after the DW/DM has been loaded successfully.

A staging area can be used for any of the following purposes.

- To gather data from different sources that will be ready to process at different times.

- To quickly load information from the operational database, freeing it up as soon as possible. All the needed transformations can then occur without interfering with the operation.
- To find changes against current DW/DM values.
- For 'data cleansing'
- To pre-calculate aggregates.
- Since the data staging structure does not need to resemble the structure of either the source or the target, the mere process of loading the staging area can be a step of the ETL (extraction, transformation, loading) process in itself.

(i) ETL-Data Extraction

Data Extraction is an activity, which pulls the data from various data sources. Most of these sources are production systems OR are used for transaction level work.

(ii) ETL-Data Transformation

The Transformation makes sure that the transaction level raw data is transformed into a form (while still being detailed) so that it can be loaded into the 'presentation/Loaded' area.

(iii) ETL-Presentation/Loaded 'Area'

This is the repository where the data is finally loaded after going through all the works of Extraction and Transformation. This becomes the ultimate source for information for various reasons ranging from queries to advanced data modeling.

(C) Meta Data

Meta Data is the data about the data. It contains all the business and technical designs, rules and locations etc. of all the data starting from the Extraction to final data usage.

(D) End User Tools and Applications

Data is ready for consumption. There is a long list of applications to which the data can be put to and the tools, which can make it happen. This includes the reporting, publishing, analysis, modeling and mining tools.

Data warehouse is a large platform, which has large number of users, data sources and data targets. Just like production systems, it has to be administered in terms of performance, timelines and availability. This also includes activity logging, data security, backing-up and archiving.

1.10. DATA WAREHOUSE ARCHITECTURE

Data warehouses and their architectures vary depending upon the specifics of an organization's situation. Three common architectures are:

- Data Warehouse Architecture (Basic)
- Data Warehouse Architecture (with a Staging Area)
- Data Warehouse Architecture (with a Staging Area and Data Marts)

1.10.1. Data Warehouse Architecture (Basic)

Simple architecture for a data warehouse is shown in Fig. 1.6. End users can directly access data derived from several source systems through the data warehouse in this type of architecture of data warehouse. This illustrates three things:

- **Data Sources:** Data sources may be operational systems, flat files, relational databases, ERP system etc.

➡ **Data Warehouse:** Data warehouse consists of metadata, summary data, and raw data

➡ **Users:** Users may use the data for analysis, reporting, mining.

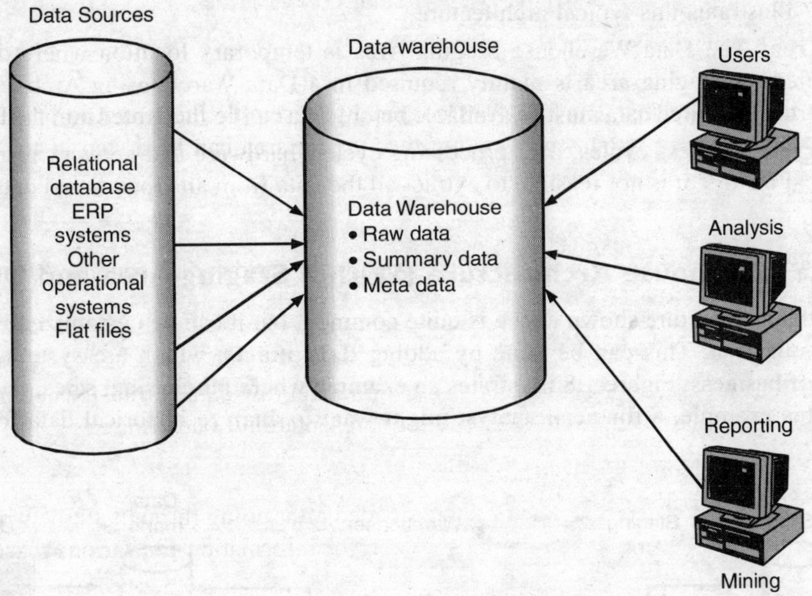

Fig. 1.6. Data Warehouse Architecture ('Basic').

1.10.2. Data Warehouse Architecture (With a Staging Area)

In the simplest (basic) architecture there is no cleaning process, but there is a need to clean the data before it will move to data warehouses. It can be done programmatically, although most data warehouses use a **staging area** instead. A staging area simplifies building summaries and general warehouse management.

➡ **Data Sources:** Operational systems and flat files, ERP systems, RDBMS.

➡ **Staging Area:** Staging area is defined as that is cleaning is done here. where data sources go before the warehouse.

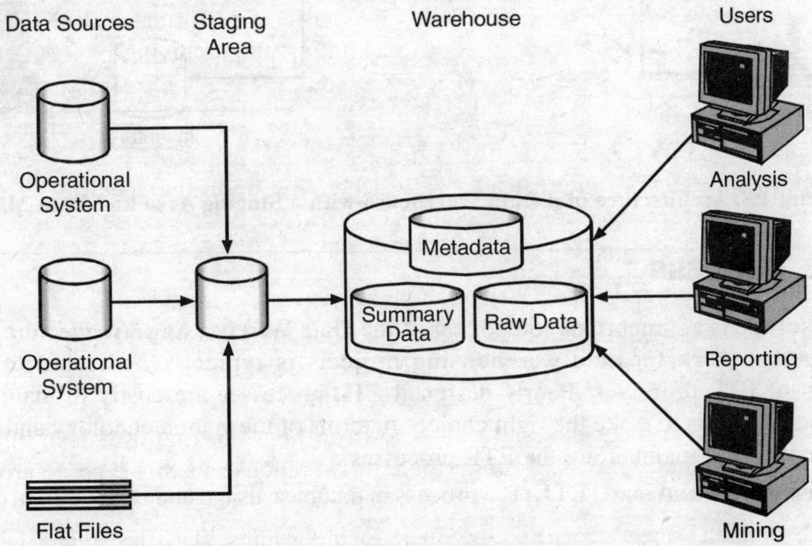

Fig. 1.7. Architecture of a Data Warehouse with a Staging Area.

- **Warehouse:** Data warehouse consists of metadata, summary data, and raw data.
- **Users:** Analysis, reporting, and mining.

Figure 1.7 illustrates this typical architecture.

Staging Area: The Data Warehouse Staging Area is temporary location where data from source systems is copied. A staging area is mainly required in a Data Warehousing Architecture for timing reasons. In short, all required data must be available before data can be integrated into the Data Warehouse.

Due to varying business cycles, data processing cycles, hardware and network resource limitations and geographical factors, it is not feasible to extract all the data from all Operational databases at exactly the same time.

1.10.3. Data Warehouse Architecture (With a Staging Area and Data Marts)

Although the architecture shown above is quite common, but it can be customize to different groups within the organization. This can be done by adding **data marts**, which are systems designed for a particular line of business. Figure 1.8 illustrates an example where purchasing, sales, and inventories are separated. In this example, a financial analyst might want to analyze historical data for purchases and sales.

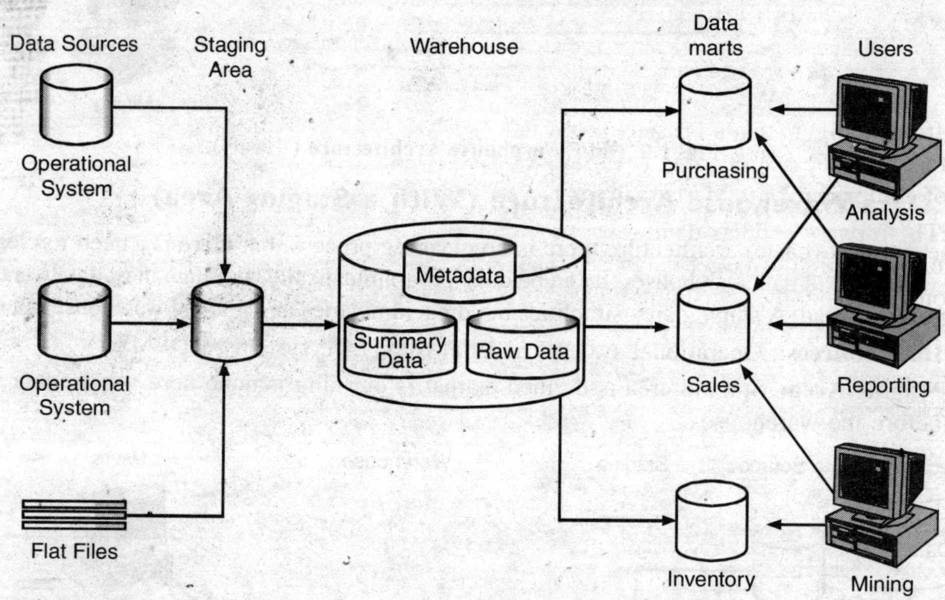

Fig. 1.8. Architecture of a Data Warehouse with a Staging Area and Data Marts

1.11. ETL PROCESS

ETL Technology is an important component of the Data Warehousing Architecture. More than half of all development work for data warehousing projects is typically dedicated to the design and implementation of ETL processes. Poorly designed ETL processes are costly to maintain, change and update, so it is critical it is to make the right choices in terms of the right technology and tools that will be used for developing and maintaining the ETL processes.

Extract, transform and load (ETL) is a process in database usage and especially in data warehousing that involves:

- Extracting data from outside sources

- ➠ Transforming it to fit operational needs (which can include quality levels)
- ➠ Loading it into the end target (database or data warehouse)

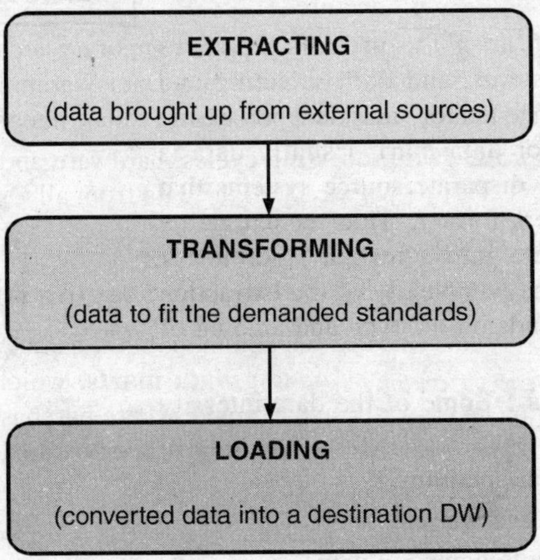

Fig. 1.9. ETL Process

Clear from the figure that it is used to copy data from Operational Applications to the Data Warehouse Staging Area, from the DW Staging Area into the Data Warehouse and finally from the Data Warehouse into a set of conformed Data Marts that are accessible by decision makers.

The ETL software extracts data, transforms values of inconsistent data, cleanses "bad" data, filters data and loads data into a target database. The scheduling of ETL jobs is critical. Should there be a failure in one ETL job, the remaining ETL jobs must respond appropriately.

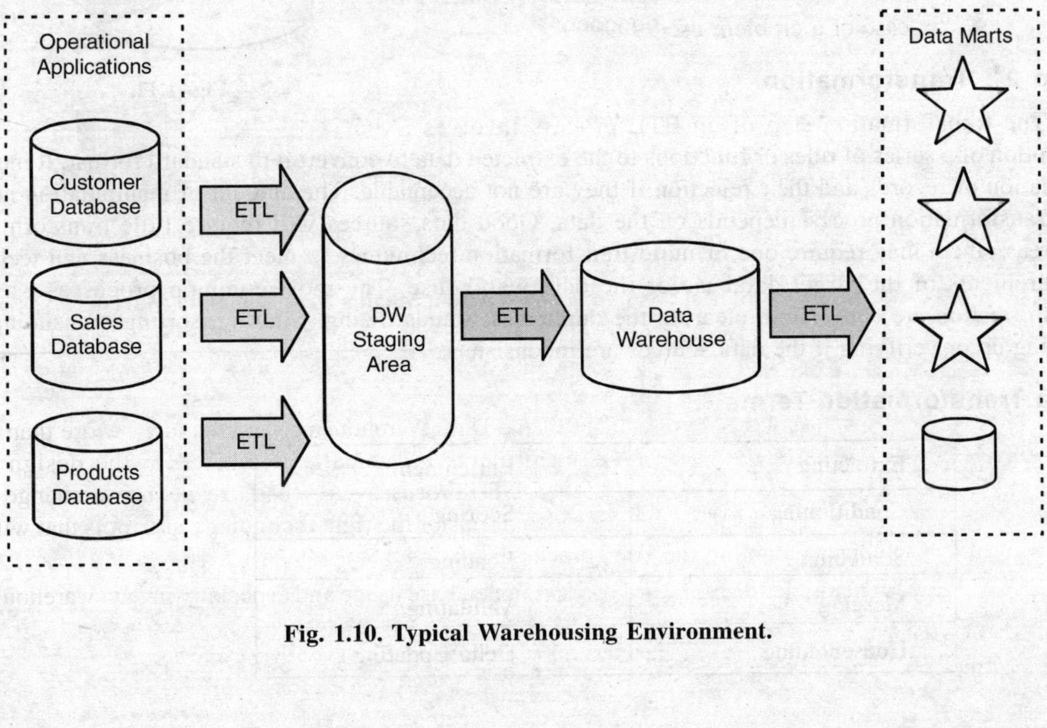

Fig. 1.10. Typical Warehousing Environment.

Three Steps in ETL Process

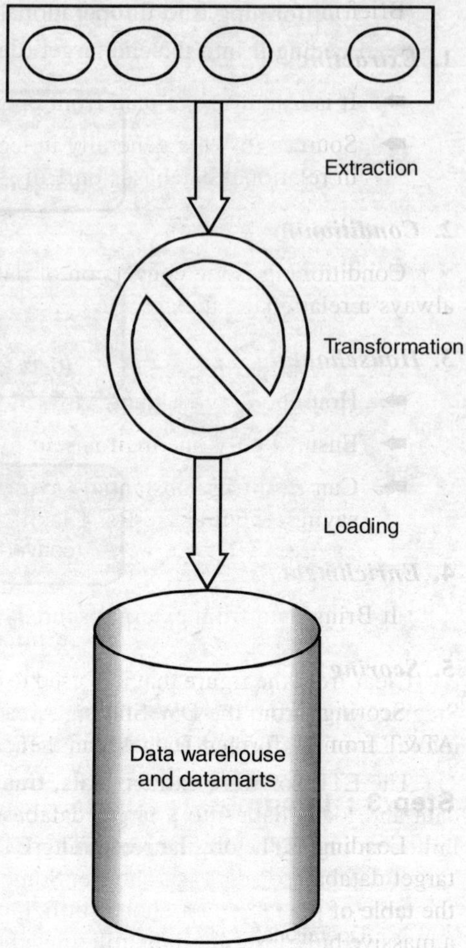

Fig. 1.11.

Step 1 : Extraction

The extraction step of an ETL process involves connecting to the source systems, and both selecting and collecting the necessary data needed for analytical processing within the data warehouse or data mart. Usually data is consolidated from numerous, disparate source systems that may store the date in a different format. Thus the extraction process must convert the data into a format suitable for transformation processing. The complexity of the extraction process may vary and it depends on the type and amount of source data.

Data Integrity Problems : Some of the data integrity problems are :

- Same person, different spellings
  Agarwal, Agrawal, Aggarwal etc...
- Multiple ways to denote company name
  Persistent Systems, PSPL, Persistent Pvt. LTD.
- Use of different names
  mumbai, bombay
- Different account numbers generated by different applications for the same customer
- Required fields left blank
- Invalid product codes collected at point of sale
  - manual entry leads to mistakes
  - "in case of a problem use 9999999"

Step 2 : Transformation

The transformation step of an ETL process involves execution of a series of rules or functions to the extracted data to convert it to standard format. It includes validation of records and their rejection if they are not acceptable. The amount of manipulation needed for transformation process depends on the data. Good data sources will require little transformation, whereas others may require one or more transformation techniques to meet the business and technical requirements of the target database or the data warehouse. The most common processes used for transformation are conversion, clearing the duplicates, standardizing, filtering, sorting, translating and looking up or verifying if the data sources are inconsistent.

Data Transformation Terms

Extracting	Enrichment
Conditioning	Scoring
Scrubbing	Loading
Merging	Validating
Householding	Delta Updating

Brief introduction of all these terms is given below:

1. ***Extracting***

- It is Capturing of data from operational source in "as is" status.
- Sources for data generally in legacy mainframes in VSAM, IMS, IDMS, DB2; more data today in relational databases on Unix.

2. ***Conditioning***

Conditioning is the conversion of data types from the source to the target data store (warehouse) — always a relational database.

3. ***Householding***

- Householding is Identifying all members of a household (living at the same address)
- Ensures only one mail is sent to a household
- Can result in substantial savings: 1 lakh catalogues at Rs. 50 each costs Rs. 50 lakhs. A 2% savings would save Rs. 1 lakh.

4. ***Enrichment***

It Bring data from external sources to augment/enrich operational data.

5. ***Scoring***

Scoring is the computation of a probability of an event. e.g., chance that a customer will defect to AT&T from MCI, chance that a customer is likely to buy a new product.

Step 3 : Loading

Loading is the last step of ETL process involves importing extracted and transformed data into a target database or data warehouse. Some load processes physically insert each record as a new row into the table of the target warehouse utilizing a SQL insert statement. Whereas other load processes include a massive bulk insert of data utilizing a bulk load routine. The SQL insert is a slower routine for imports of data, but does allow for integrity checking with every record. The bulk load routine may be faster for loads of large amounts of data, but does not allow for integrity check upon load of each individual record.

Issues

- Huge volumes of data to be loaded
- Small time window available when warehouse can be taken off line (usually nights)
- When to build index and summary tables
- Allow system administrators to monitor, cancel, resume, change load rates
- Recover gracefully — restart after failure from where you were and without loss of data integrity.

1.12. ADVANTAGES AND DISADVANTAGES OF DATA WAREHOUSE

Advantages of the Data warehouse are :

- Complex queries can be executed easily and efficiently because query execution does not involve data translation and communication with remote sources.
- End users can use a single data model and query language.
- Easily Facilitate Process change.

- Effective decision making can be made
- System design becomes simpler. For example, there is no need to perform query optimization over heterogeneous sources.
- Information sources may be unreliable and may purge data. On the other hand, information at the warehouse is under the control of the warehouse users; it can be stored safely and reliably for as long as necessary.
- To manage customer and business partner relationship effectively.

Disadvantages

- Additional Hardware/Software needed.
- Maintenance is causing a major set back for data warehousing

Summary

- **Data warehousing is defined as a**
  1. Technique for assembling and managing data from various sources for the purpose of answering business questions. Thus making decisions that were not previous possible.
  2. A decision support database maintained separately from the organization's operational database.
- A data warehouse is a blend of many technologies as it takes data from the operational systems and from the outside sources like reports, journals etc, removes the inconsistencies transform the data and finally loads in it in the formats suitable for easy access for decision making.
- A data warehouse is a
  1. subject-oriented
  2. integrated
  3. time-varying
  4. non-volatile
  5. collection of data that is used primarily in organizational decision making.
- The advantages of data warehousing includes increased productivity, better decisions, liability management, better CRM.
- Before implementing data warehouse the ETL process is necessary. ETL is extraction, transformation and loading.
- A data mart is defined as an implementation of a data warehouse with small and more tightly restricted scope of data and data warehouse functions, serving a single department or part of an organization.
- Advantage of Data Mart is that it provides high support for end-user work groups, departments or small organization. Data Marts are usually linked with DSS tools and provide high functionality with these tools. Restricted scope means its usability and its life cycle not in size. Data Mart is tool independent.
- Disadvantage of Data Mart. High cost and reduced data consistency, in long run. For each data mart model there is no common consolidated source for the various mart. Cross data mart issue, such as consistency, integration are difficult to resolved. Not be static.

- A Data Mart is not Data Warehouse A data mart is not just small data warehouse A collection of Data mart is not a Data Warehouse A data Warehouse is not a DSS.
- A Management information system (MIS) also known as virtual ware house, is defined as a query and reporting system that directly performs aggregation and summarization against operational data.
- **Characteristics of MIS:** Direct access to source data. An MIS focused on current data. Facing difficulties when deal with external data and with unstructured data.

Review Questions

1. Comment on the statement "data warehouse is an environment and not a product".
2. Define data warehousing with its advantages and disadvantages.
3. Explain ETL process in detail.
4. Explain the factors which lead to growth of data warehouse.
5. Explain the difference between DBMS vs. data warehouse.
6. Write a short note on benefits of data warehouse.
7. What do you mean by "basic" architecture of DW.
8. Why staging area is used in data warehouse ?
9. What is the function of data marts ?
10. Differentiate between OLTP and DW.

2

Chapter

Data Warehouse: Features and Architecture

INSIDE THIS CHAPTER

2.1. Features of Data Warehouse **2.2.** Data Granularity **2.3.** Metadata **2.4.** Architecture of Data Warehouse **2.5.** Data Warehouse Models **2.6.** Data Marts.

In this chapter we will cover the features and the architecture of data warehouse. The basics of Data Warehouse have been already covered in the last chapter. Five defining **features of Data Warehouse** are subject-oriented, integrated, time-variant, non-volatile and data granularity. Different types of data warehouse architecture is also discussed in this chapter.

2.1. FEATURES OF DATA WAREHOUSE

The most significant differences between the corporate data warehouse and other enterprise-wide systems involve the methods used to store and manage data.

A data warehouse has significantly different features from other enterprise-wide systems, particularly in how data is stored, managed and manipulated. After this topic it will be easy to define data warehouse features and differentiate with operational database by analyzing these points.

There are five key characteristics which separate the data warehouse from other major operational systems:

- **Subject Orientation:** Subject Orientation means that data is organized by subject.
- **Integration:** Consistency of defining parameters. Integration deals with consistency of defining parameters.
- **Non-volatility:** Nonvolatility means data storage medium must be stable.
- **Time-variance:** Time variance means timeliness of data and access terms.
- **Data granularity:** Data granularity means that details of data are kept at low level.

The detailed description of all these features are described here.

NOTE

*In Data Warehousing, the **Operational Database** is one which is accessed by an Operational System to carry out regular operations of an organization. Operational Databases usually use an OLTP Database which is optimized for faster transaction processing: Inserting, Deleting, and Updating data. On the other side, **Data Warehouses** use an OLAP Database (Online Analytical Processing) which is optimized for faster queries. An Operational Database is usually put on a joined machine from the Data Warehouse to increase performance.*

2.1.1. Subject-oriented

In operational database, we store data by individual applications. In the data sets for an order processing application, we keep the data for that particular application. These data sets provide the data for all the functions for entering orders, checking stock, verifying customer's credit, and assigning the order for shipment. But these data sets contain only the data that is needed for those functions relating to this particular application. We will have some data sets containing data about individual orders, customers, stock status, and detailed transactions, but all of these are structured around the processing of orders.

In the data warehouse, data is stored by subjects, not by applications. If data is stored by business subjects, what are business subjects? Business subjects differ from enterprise to enterprise. These are the subjects critical for the enterprise. For a manufacturing company, sales, shipments, and inventory are critical business subjects. For a retail store, sales at the check-out counter are a critical subject.

2.1.2. Integrated

For proper decision making, there is a need to pull together all the relevant data from the various applications. The data in the data warehouse comes from several operational systems. Source data are in different databases, files, and data segments. These are dissimilar applications, so the operational platforms and operating systems could be different. The file layouts, character code representations, and field naming conventions all could be different.

In addition to data from internal operational systems, for many enterprises, data from outside sources is likely to be very important. Companies such as Metro Mail, and IRI specialize in providing vital data on a regular basis. The data warehouse may need data from such sources. This is one more variation in the mix of source data for a data warehouse.

Before the data from various disparate sources can be usefully stored in a data warehouse, there is a need to remove the inconsistencies. The various data elements and meanings of data names in each source application must be standardised. Before moving the data into the data warehouse, the data must go through a process of transformation, consolidation, and integration of the source data.

2.1.3. Time-variant

For an operational system, the stored data contains the current values. In an accounts receivable system, the balance is the current outstanding balance in the customer's account. In an order entry system, the status of an order is the current status of the order. In a consumer loans application, the balance amount owed by the customer is the current amount. Of course, we store some past transactions in operational systems, but, essentially, operational systems reflect current information because these systems support day-to-day current operations.

On the other hand, the data in the data warehouse is meant for analysis and decision making. If a user is looking at the buying pattern of a specific customer, the user needs data not only about the current purchase, but on the past purchases as well. When a user wants to find out the reason for the drop in sales in the South West division, the user needs all the sales data for that division over a period extending back in time. When an analyst in a grocery chain wants to promote two or more products together, that analyst wants sales of the selected products over a number of past quarters.

A data warehouse, because of the vary nature of its purpose, has to contain historical data, not just current values. Data is stored as snapshots over past and current periods. Every data structure in the data warehouse contains the time element. We can also find historical snapshots of the operational data in the data warehouse. This aspect of the data warehouse is quite significant for both the design and the implementation phases.

2.1.4. Nonvolatile

Data extracted from the various operational systems and relevant data obtained from outside sources are transformed, integrated, and stored in the data warehouse. The data in the data warehouse is not intended to run the day-to-day business. When we want to process the next order received from a customer, we do not look into the data warehouse to find the current stock status. The operational order entry application is meant for that purpose. In the data warehouse, extracted stock status data as snapshots over time is kept. The data warehouse need not be updated every time when we process a single order.

Data from the operational systems are moved into the data warehouse at specific intervals. Depending on the requirements of the business, these data movements take place twice a day, once a day, once a week, or once in two weeks. In fact, in a typical data warehouse, data movements to different data sets may take place at different frequencies. The changes to the attributes of the products may be moved once a week. Any revisions to geographical setup may be moved once a month. The units of sales may be moved once a day. We can plan and schedule the data movements or data loads based on the requirements of your users.

NOTE ▶

OLTP Systems are used to "run" a business and The Data Warehouse helps to "optimize" the business

2.1.5. Data Granularity

Granularity simply means the level of detail. High granularity refers to data that is at or near the transaction level. Data that is at the transaction level is usually referred to as atomic level data. Low granularity refers to data that is summarized or aggregated, usually from the atomic level data. Summarized data can be lightly summarized as in daily or weekly summaries or highly summarized data such as yearly averages and totals. Data granularity is explained in detail in next topic.

2.2. DATA GRANULARITY

In an operational system, data is usually kept at the lowest level of detail. For example in a point-of-sale system for a grocery store, the units of sale are captured and stored at the level of units of a product per transaction at the check-out counter. In an order entry system, the quantity ordered is captured and stored at the level of units of a product per order received from the customer. Whenever summary data is needed, the individual transactions are added. If we are looking for units of a product ordered this month, all the orders entered for the entire month for that product are read out and add up. Usually summary data in an operational system is not kept.

When a user queries the data warehouse for analysis, he or she usually starts by looking at summary data. The user may start with total sale units of a product in an entire region. Then the user may want to look at the breakdown by states in the region. The next step may be the examination of sale units by the next level of individual stores. Frequently, the analysis begins at a high level and moves down to lower levels of detail.

In a data warehouse, it is efficient to keep data summarized at different levels. Depending on the query, you can then go to the particular level of detail and satisfy the query. Data granularity, the last *features of data warehouse*, in a data warehouse refers to the level of detail. The lower the level of detail, the finer the data granularity. That is, if data is kept in the lowest level of detail, lots of data needs to be store in the data warehouse. Thus granularity levels are to be decided based on the data types and the expected system performance for queries.

NOTE

The choice of granularity is an important design issue because:

1. *The lower the level of granularity, the larger is the amount of data stored in the warehouse.*
2. *The higher the level of granularity, the lesser is the level of detail for which queries can be answered.*
3. *The choice of granularity is a tradeoff between the volume of data and level of query detail.*

2.2.1. Advantages and Disadvantages of Data Granularity

The data granularity is the most important design issue in data warehouse. Granularity Controls the volume of data in the DW.

Advantages

- **Reusability and Flexibility:** DW provides an invaluable foundation for many different types of DSS processing. Organizations may build a DW for one purpose and then discover that it can be used for many other kinds of DSS processing. Although infrastructure for the DW is expensive and difficult to build, it has to be built only once. Granular data is the key to reusability. Different departments are able to look at the data as they wish to see it. A related benefit is the ability to reconcile data, if needed. Once there is a single foundation (granular data) on which everybody relies, if there is a need to explain a discrepancy in analyses between two or more departments, then reconciliation is relatively simple. Another related benefit is flexibility. Any department can alter how it looks at data.
- Data granularity allows us to issue more versatile queries. The lower the level of granularity, the more versatile the query can be issued.
- Future unknown requirements can be accommodated easily on the foundation of granular data
- ETL process is simplified as no summarization is required.
- Granular data is an ideal source of data for the down stream Data Mining applications.
- Dimension of any granularity can be added to an existing star schama (explained later).
- In the bottom-up approach, the integration of data marts becomes easy if the data marts are having the most granular data (supermarts). In the top-down approach, each data mart that sources its data from the DW may have its own unique requirements. Any such requirement can be handled by the most granular data.

Disadvantages

The disadvantage of Data Granularity is that sometimes data volume could become unmanageable. For example, the web log data generated by the web based eBusiness environment (often called clickstream data) is at a very low level of granularity. It must be edited, filtered, and summarized.

2.3. METADATA [MDU May 2012]

The general meaning of Metadata is data about data. But it is an ambiguous term which is used for two fundamentally different concepts (types). Although the expression "data about data" is often used, it does not apply to both in the same way.

1. **Structural Metadata :** Structured metadata is the design and specification of data structures, and it cannot be about data, because at design time the application contains no data. In this case the correct description would be "data about the containers of data".
2. **Descriptive Metadata :** Descriptive metadata is about individual instances of application data, the data content. In this case, it would be "data about data content" or "content about content" thus met content.

   The success of Data warehouse depends on Metadata.

2.3.1. What is Metadata?

Before going in to more details of metadata , we will first understand what is Metadata and why it is important in DW environment. Better end user data access and analysis tools can help users to figure out that how to get information they need, out of the data warehouse, but only good, easily accessible metadata can help them figure out what is available in the data warehouse and how to ask for it. Thus the success of DW depends on Metadata.

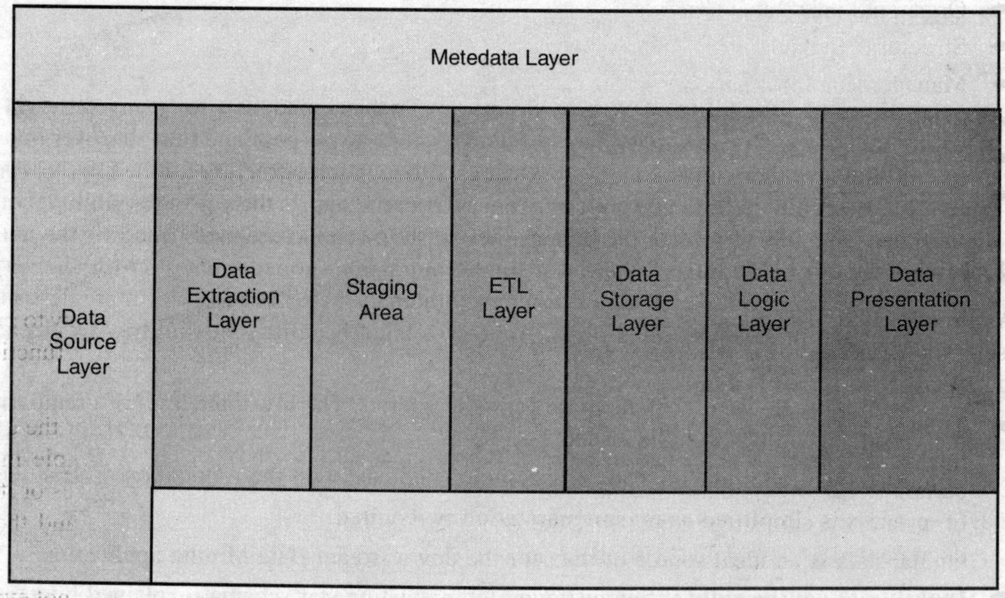

Fig. 2.1. Metadata Layer.

There are three main categories of Metadata according to Kimball et al. these are technical metadata, business metadata and process metadata. Technical metadata is primarily definitional while business metadata and process metadata are primarily descriptive. Sometimes the categories may overlap.

- **Technical Metadata** defines the objects and processes in a DW/BI system, as seen from a technical point of view. The technical metadata includes the system metadata which defines the data structures such as: Tables, fields, data types, indexes and partitions in the relational engine, and databases, dimensions, measures, and data mining models. Technical metadata defines the data model and the way it is displayed for the users, with the reports, schedules, distribution lists and user security rights.

- **Business Metadata** is content from the data warehouse that is described in more user-friendly terms. The business metadata tells us about what data we have, where it comes from, what it means and what its relationship is to other data in the data warehouse. Business metadata may also serves as documentation for the Data Warehouse (DW)/Business Information (BI) system. Users who browse the data warehouse are primarily viewing the business metadata.
- **Process Metadata** is used to describe the results of various operations in the data warehouse. Within the ETL process all key data from tasks are logged on execution. This includes start time, end time, CPU seconds used, disk reads, disk writes and rows processed. When troubleshooting the ETL or query process, this sort of data becomes valuable. Process metadata is the fact measurement when building and using a DW/BI system. Process metadata is in interest of business people who can use the data to identify the users of their products, which products they are using and what level of service they are receiving.

A metadata does not gives just the description of the entity but also gives the other details explaining the syntax and semantics of the data elements. Metadata described all the important aspects of data in data warehouse fully and precisely to help the users and the developers of the data warehouse. A metadata contains the information about the data warehouse system and this includes:

- Content Information
- Organizational Behaviour
- Structural Behaviour
- Management Information
- Scheduling Information
- Contact Information
- Technical Information

2.3.2. Need of Meta Data

As already said that metadata stores the data about the data in data warehouse. It is the key to providing the users and developers the roadmap to the information in the warehouse. The need or functions that metadata performs in a data warehouse are:

- **To maintain the data quality :** Most organizations have to live with the reality of the quality of legacy and external data due to poor validation rules, overuse of fields or simple misuse of certain fields. To maintain the data quality is very important during the early phases or iterations of a data warehousing project that the quality of the data should be analyzed and the results documented as associated with meta data.
- **Inability to get the right information :** Without the proper meta data, users may not even know certain information already exists in the data mart or a data warehouse.
- **Productivity loss :** Users are not able to get to the right information in the right way quickly resulting in productivity loss
- **Unpredictable performance :** Lack of information on query estimates impairs the ability to schedule long-running queries to off-peak hours. This can result in unpredictable performance. To avoid this problem, meta data should contain estimated time and expected number of resultant rows of at least the seemingly complex and long queries. It is understandable that in a DW environment it will be almost impossible to estimate all the ad hoc queries. However, building a simple historical file about various queries can go a long way in alleviating this problem.

➡ **Difficult to upgrade and enhance :** Data warehouse developers will not be able to maintain, enhance or upgrade a data mart or a data warehouse without a good knowledge of what is in the current version of the warehouse. This is especially true with current trends toward implementing quick data marts. If done without proper meta data architecture, users will get inconsistent and conflicting information from data marts.

2.3.3. Characteristics of Metadata Architecture

Metadata architecture should have the following characteristics:

➡ **Mandatory :** Meta data has been very important even for OLTP systems. However, for OLTP systems, most of the meta data was required by the IT community consisting of programmers and analysts. It never acquired the importance it has with the onset of data warehousing where most of the meta data is required by end users. They would be totally lost if they didn't know what was available in the warehouse, what it means and how to access it. Data warehouse users should be provided not just accurate meta data but accurate contextual meta data. As described earlier, without accurate and up-to-date contextual meta data information, they can get misleading and ambiguous information, which can lead to wrong decisions. Thus meta data architecture for a data warehousing project should not be an afterthought, but a mandatory and well planned part of the overall architecture.

➡ **Open :** (ability to work and get information from various categories of tools). It is a key requirement for the meta data to be up to date. If the meta data repository is proprietary and does not interchange information with other components of the data warehouse architecture, the meta data update process will be tedious, resulting in out-of-date meta data. As shown in the Figure 2, the meta data tool should be able to work seamlessly with various sources of meta data:

* CASE tools;
* Applications for the source data, both internal and external;
* Database engines, whether relational or multidimensional;
* Data extraction, transformation and cleansing tools;
* DW monitors; and
* Front-end tools.

➡ **Easy to synchronize and update :** If the meta data manager requires jumping through a lot of hoops to keep meta data up to date, it is a sure recipe for failure. The architecture should facilitate automatic updates with sources of meta data wherever possible. There should be clear cut standard procedures for updating and synchronizing meta data.

	Business Analysis	Data Analysis & Mapping	Architecture Design	Development & Implementation	Maintenance & Support
CASE	✔	✔	✔		✔
Database Engines				✔	✔
Data Extraction, Cleansing & Transformation		✔	✔	✔	✔
Front-End Tools				✔	
Data Warehouse Administration Tools					✔
Legacy Source Applications	✔	✔			✔

Fig. 2.2. Meta Data Source

The active meta data manager would be the core component of the meta data architecture and would ideally consist of the following components:

- **Meta Data Capture :** It means initially capture of meta data from a variety of sources.
- **Meta Data Synchronizer :** M-data synchronizer processes to keep meta data up to date.
- **Meta Data Search Engine :** This is the front end for the users to search and access meta data.
- **Meta Data Results Manager :** This processes the results of the meta data search and allow the user to make an appropriate selection.
- **Meta Data Alerter :** As part of a push technology, this notifies the subscribers about any new changes to the meta data contents depending on the user profile.
- **Meta Data Query Trigger :** M-Data query trigger, triggers an appropriate query tool to get data from a data warehouse or any other source based on the selection made by the user in the meta data results manager.

2.3.4. Metadata Lifecycle

Metadata life cycle consists of following steps :

- **Collection:** Collection means to identify metadata and capture into repository; automate.
- **Maintenance:** Maintenance is to put in place the processes to synchronize metadata automatically with changing data architecture.
- **Deployment:** Deployment means to provide metadata to users in the right form and with the right tools; match metadata offered to specific needs of each audience.

2.3.5. Metadata Repository

These are the specialized databases that are designed to maintain metadata, together with tools and interfaces that allow a company to collect and distribute its metadata. The metadata repository requirements are :

- Logically Common
- Open
- Extensible (as explained above)

Multiple repository is an extension of simple repository as clear from the diagram.

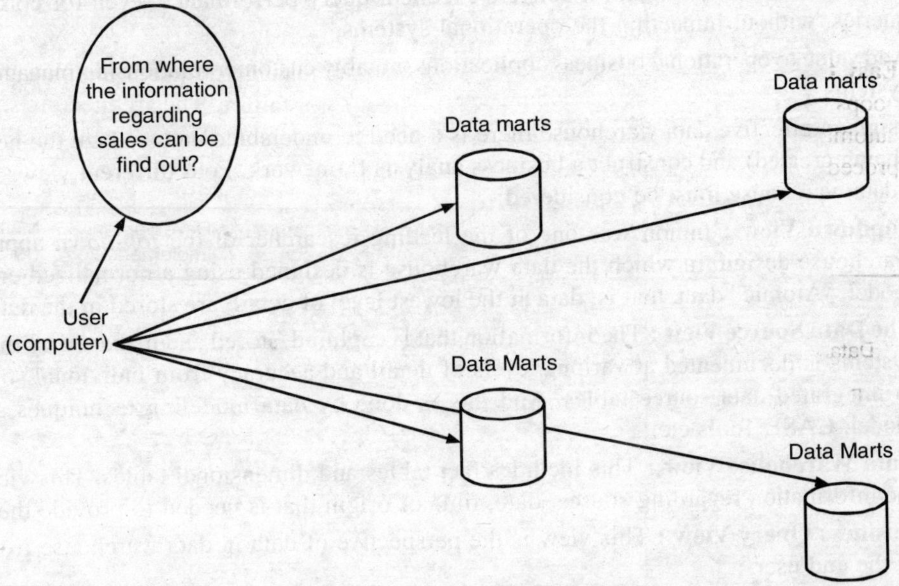

Fig. 2.3. Metadata Repository.

But there are some challenges regarding Metadata:

Since the source systems contain the data in different formats, having different standards, definitions, attributes, values etc. Therefore it becomes a cumbersome task to unify the metadata. Also the passage of data from one process to other is not easy.

2.4. ARCHITECTURE OF DATA WAREHOUSE

Architecture is defined as the logical and physical foundation on which the system is build. Although the architectures of data warehouse have been already studied in 1st chapter, but here the architecture is described in detail.

The architecture of a data warehouse includes everything that is needed to prepare the data and store it and finally all the means for delivering the information from the data warhouse to the users. The data warehousing architecture is wide, complex, and expensive. The steps for the design and construction of data warehouses are explained first.

2.4.1. Design of Data Warehouse

The Advantages of Data Warehouse

A data warehouse maintains a copy of information from the source transaction systems. This architectural complexity provides the opportunity to:

- Maintain data history, even if the source transaction systems do not.
- Integrate data from multiple source systems, enabling a central view across the enterprise. This benefit is always valuable, but particularly so when the organization has grown by merger.
- Improve data quality, by providing consistent codes and descriptions, flagging or even fixing bad data.
- Present the organization's information consistently.
- Provide a single common data model for all data of interest regardless of the data's source
- Restructure the data so that it makes sense to the business users.
- Restructure the data so that it delivers excellent query performance, even for complex analytic queries, without impacting the operational systems.
- Add value to operational business applications, notably customer relationship management (CRM) systems.

To design an effective data warehouse there is a need to understand and analyze the needs (why the system is being created) and construct a business analysis framework. Four different views regarding the design of data warehouse must be considered:

1. **Topdown View :** Inmon was one of the leading researcher of the *top-down* approach to data warehouse design, in which the data warehouse is designed using a normalized enterprise data model. "Atomic" data, that is, data at the lowest level of detail, are stored in the data warehouse.
2. **The Data Source View :** The information that is captured, stored and managed by the operational systems is documented at various levels of detail and accuracy from individual source systems to integrated data source tables. And this is done by data modeling techniques such as : ER model, CASE Tools etc.
3. **Data Warehouse View :** This includes fact tables and dimensional tables. This view also gives the information regarding source, date, time of origin that is needed to provide the history.
4. **Business Query View :** This view is the perspective of data in data warehouse from viewpoint of the end user.

Thus the goal of designing the data warehouse is to serve the users and it should be designed to satisfy the following requirements :

(1) Data warehouse must provide user friendly environment, allows the users to do query processing without interfering with the operational systems.

(2) Data warehouse must provide a central repository of consistent, accurate, cleaned data, and there should be no inconsistencies and anamolies in the data and must respond to complex queries quickly and efficiently.

The basics of data warehouse architecture is explained earlier but mainly the data warehouse architecture is studied under two tier data warehouse architecture and three tier data warehouse architecture

2.4.2. Two Tier Data Warehouse Architecture—Independent Data Marts

As clear from the name two tier architecture consists of client and the server. Client at the front end and server at the back end. Client machine deals with

* GUI/Presentation logic
* Query specification
* Data analysis
* Report formatting
* Summarizing
* Data access

and the DW Server deals with :

* Data logic
* Data services
* Meta data
* File services

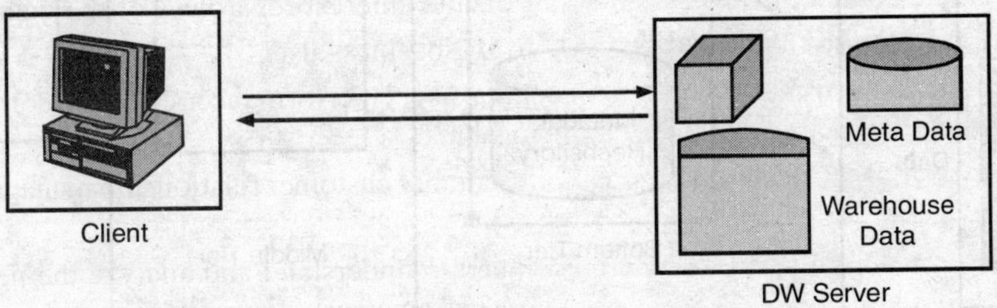

Fig. 2.4. Two Tier Architecture of Data Warehouse.

Two tier architecture consists only client and servers and thus independent of data marts.

The advantages of this architecture is that minimal investment in additional hardware and software but the disadvantage of this is that it is not scalable and can't support a large number of online end users without additional software modifications.

2.4.3. Three Tier Architecture

[MDU Dec. 2011; May 2012]

Due to limitations of two tier data warehouse architectures. Data warehouse adopt a three tier architecture. The components of this three tier architecture are :

1. Bottom Tier (Data warehouse server)
2. Middle Tier (OLAP server)
3. Top Tier (Front end tools).

Bottom Tier or Data Warehouse Server

Bottom tier is a warehouse database server that is a relational database system in which data is fed using Back end tools and utilities and data is extracted from bottom tier using programs called gateways. This data warehouse server also contains Meta data repositories

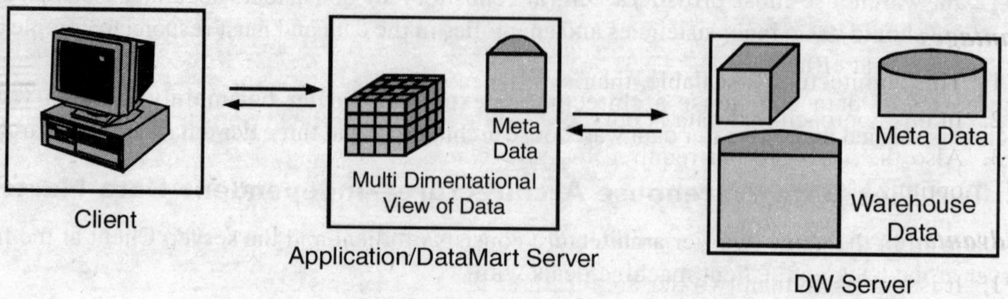

Fig. 2.5. Three Tier Architecture of Data Warehouse.

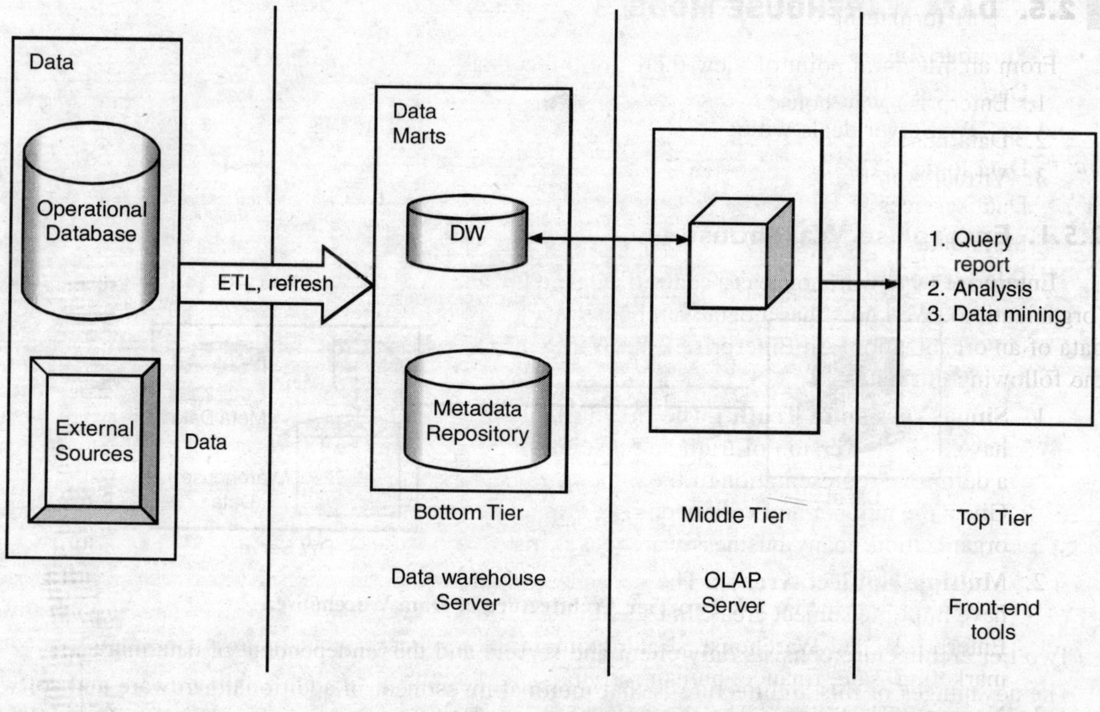

Fig. 2.6. Detailed View of Three-Tier Architecture of Data Warehouse

Middle Tier or OLAP Server

The middle tier is an OLAP server that is typically implemented using either

(1) a relational OLAP (ROLAP) model, that is, an extended relational DBMS that maps operations on multidimensional data to standard relational operations; or

(2) a multidimensional OLAP (MOLAP) model, that is, a special-purpose server that directly implements multidimensional data and operations.

Top Tier or Font End Tool Analysis

The top tier is a front-end client layer, which contains query and reporting tools, analysis tools, and/or data mining tools.

Advantages and Disadvantages of three tier architecture are :

Advantages

1. This architecture is scalable, than two tier model.
2. In this approach that data is put closer to the end users, thus response time is better.
3. Also the users are not required to know where or how the data is stored thus performance is optimized.

Disadvantages

1. It is expensive than two tier architecture.
2. Local data administration is a necessity as the data design needs to be controlled and optimized for the queries.

2.5. DATA WAREHOUSE MODELS

From architectural point of view there are three data warehouse models

1. Enterprise warehouse.
2. Data mart.
3. Virtual warehouse.

2.5.1. Enterprise Warehouse

Enterprise Data Warehouse is a centralized warehouse which provides service for the entire enterprise (organization). We know that a data warehouse is a large repository of historical and current transaction data of an organization. An Enterprise Data Warehouse is a specialized data warehouse which may have the following attributes.

1. **Single Version of Truth :** The first attribute of an Enterprise Data Warehouse is that it should have a single version of truth and that entire goal of the warehouse's design is to come up with a definitive representation of the organization's business data as well as the corresponding rules. Given the number and variety of systems and lots of company data that exist within any business organization, many business warehouses may not qualify as an Enterprise Data Warehouse.
2. **Multiple Subject Areas :** The second attribute is that an Enterprise Data Warehouse should have multiple subject areas. In order to have a unified version of the truth for an organization, an Enterprise Data Warehouse should contain all subject areas related to the enterprise such as marketing, sale, finance, human resource and others.
3. **Normalized Design :** The third attribute is that an Enterprise Data Warehouse should have a normalized design. This may be an arguable attribute as both normalized and de-normalized databases have their own advantages for a data warehouse. In fact, may data warehouse designers have used denormalized models such as star or snowflake schemas for implementing data marts. But many also go for normalized databases for an Enterprise Data Warehouse in the consideration of flexibility first and performance second.
4. **Adapted to Critical Environment :** The fourth attribute is that an Enterprise Data Warehouse should be implemented as a Mission-Critical Environment. The entire underlying infrastructure should be able to handle any unforeseen critical conditions because failure in the data warehouse means stoppage of the business operation and loss of income and revenue. An Enterprise Data

Warehouse should have high availability features such as online parameter or database structural changes, business continuance such as failover and disaster recovery features and security features.

5. **Scalable :** Finally an Enterprise Data Warehouse should be scalable across several dimensions. It should expect that a company's main objective is to grow and that the warehouse should be able to handle the growth of data as well as the growing complexities of processes which will come together with the evolution of the business enterprise.

An enterprise data warehouse may be implemented on traditional mainframes or parallel architecture platforms.

2.5.2. Virtual Data Warehouse (VDW) [MDU Dec. 2011]

VDW is an acronym for virtual datawarehouse. It provides a way of accessing data from whatever storage location it resides in, and provide a semantic map that allows the end user to view it as 'virual-ized' as to it's specific location

Advantages of Virtual Warehouse

1. **Faster time to market :** Data warehouse focus on end-user requirements to enable immediate access where you can.
2. **Complementary to physical data warehouse implementations :** It permits faster prototype implementation, and helps clarify both the type of information and the access requirements that the warehouse efforts will need to support. It provides an additional real-time data view.
3. **Open and flexible architecture :** An open environment using industry standards makes it easier to address new data sources, and potentially provides a wider selection of tools for the end user.
4. **Reduced maintenance :** By providing access to the data without burdening the user with where or how the data is stored (transparency layer), the infrastructure can be changed, without the user being aware of the changes

2.6. DATA MARTS

The data mart is a subset of the data warehouse that is usually oriented to a specific purpose or major data subject. The concept of a data mart can apply to any data whether they are operational data, evaluational data, spatial data, or metadata.

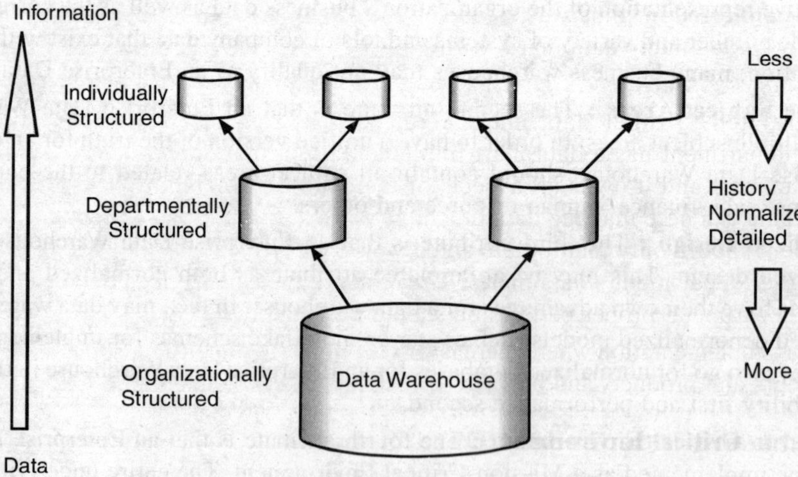

Fig. 2.7. From the Data Warehouse to Data Marts.

A data mart is a repository of a business organization's data implemented to answer very specific questions for a specific group of data consumers such as organizational divisions of marketing, sales, operations, collections and others. A data mart is typically established as one dimensional model or star schema which is composed of a fact table and multi-dimensional table.(will be explained in next chapters)

Reasons for Creating a Data Mart

There are certain reasons why to create data marts. These are :

- The frequently needed data can be easily accessed with the help of data marts.
- Data marts creates collective view by a group of users.
- End-user response time is improved.
- Ease of creation of data marts.
- The cost of creating data marts is less as compared to implementry a ful data warehouse.
- Potential users are more clearly defined than in a full data warehouse.
- Data marts contains only business essential data and is less cluttered.

2.6.1. Similarities and Dissimilarities Between DW and Data Marts

[MDU May 2012]

In comparison, a data warehouse is also a repository of organizational data implemented as a single repository serving enterprise wide data across many if not all subject areas. The data warehouse is the authoritative repository at atomic level of all fact and dimensional data.

Despite some arguments on the similarity or difference between a data mart and a data warehouse, still data mart is considered as specialized version of a data warehouse.

The data mart, like the data warehouse, can also provide a picture of a business organization's data and help the organizational staff in formulating strategies based on the aggregated data and statistical analysis of industry trends and patterns as well as part business experiences.

The most notable difference of a data mart from a data warehouse is that the data mart is created based on a very specific and predefined purpose and need for a grouping of certain data. A data mart is configured such that it makes access to relevant information in a specific area very easy and fast.

Within a single business organization, there can more than one data mart. Each of these data marts is relevant or connected in some way to one or more business units that its design was intended for. The relationship among many data marts within a single company may or may not involve inter-dependency.

They may be related to other data marts if they were designed using conformed facts of dimensions. If one department has a data mart implementation, that department is considered to be the owner of the data mart and it owns all aspects of the data mart including the software, hardware and the data itself. This can help manage data in a huge company by having a modularization method such that a department should only manipulate and develop its own data as they see if without having to alter data from other department's data marts. Then other departments need data from the data mart owned by a certain department, proper permission should be asked first.

In other data mart implementation where there is strict conformed dimension, some shared dimensions exist such as customers and products and business ownership will no longer apply.

2.6.2. Dependent and Independent Data Marts

There are two basic types of data marts: dependent and independent. The categorization is based primarily on the data source that feeds the data mart.

1. Dependent Data Marts : Dependent data marts draw data from a central data warehouse that has already been created. (Fig. 2.8)

2. Independent Data Marts : Independent data marts, in contrast, are standalone systems built by drawing data directly from operational or external sources of data, or both. (Fig. 2.9)

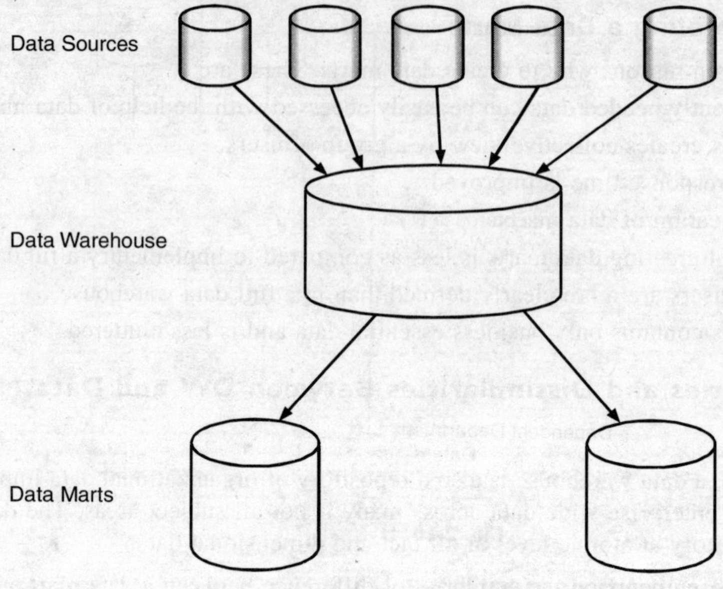

Fig. 2.8. Dependent Data Marts.

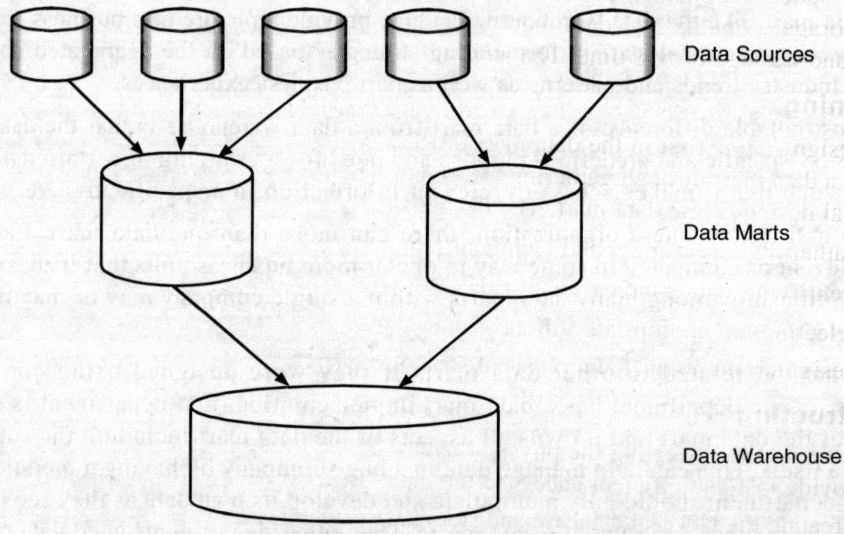

Fig. 2.9. Independent Data Marts.

Other than there two categories one more category exist that is called "Hybrid Data Marts".

3. Hybrid Data Marts : A hybrid data mart allows you to combine input from sources other than a data warehouse. This could be useful for many situations, especially when ad hoc integration is needed, such as after a new group or product is added to the organization shown in Fig. 2.10 illustrates a hybrid data mart.

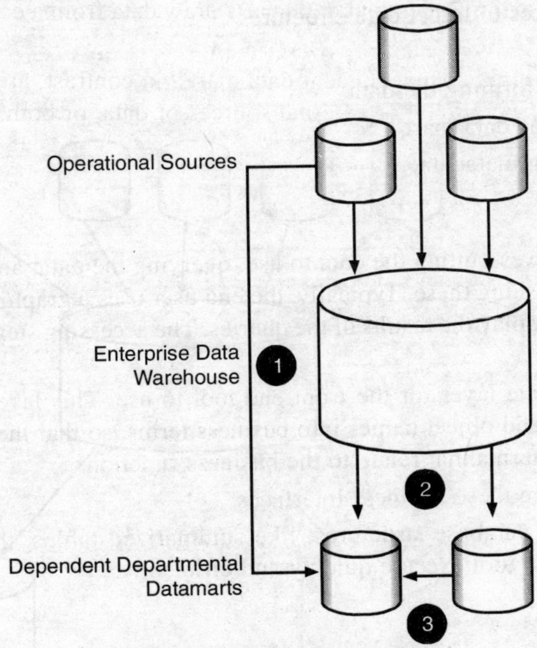

Fig. 2.10. Hybrid Data Mart.

2.6.3. The Steps in Implementing a Data Mart

Simply stated, the major steps in implementing a data mart are to design the schema, construct the physical storage, populate the data mart with data from source systems, access it to make informed decisions, and manage it over time. The steps are

1. Designing

The design step is first in the data mart process. This step covers all of the tasks from initiating the request for a data mart through gathering information about the requirements, and developing the logical and physical design of the data mart. The design step involves the following tasks:

- Gathering the business and technical requirements
- Identifying data sources
- Selecting the appropriate subset of data
- Designing the logical and physical structure of the data mart

2. Constructing

This step includes creating the physical database and the logical structures associated with the data mart to provide fast and efficient access to the data. This step involves the following tasks:

- Creating the physical database and storage structures, such as tablespaces, associated with the data mart
- Creating the schema objects, such as tables and indexes defined in the design step
- Determining how best to set up the tables and the access structures.

3. Populating

The populating step covers all of the tasks related to getting the data from the source, cleaning it up, modifying it to the right format and level of detail, and moving it into the data mart. More formally stated, the populating step involves the following tasks:

- Mapping data sources to target data structures
- Extracting data
- Cleansing and transforming the data
- Loading data into the data mart
- Creating and storing metadata

4. Accessing

The accessing step involves putting the data to use: querying the data, analyzing it, creating reports, charts, and graphs, and publishing these. Typically, the end user uses a graphical front-end tool to submit queries to the database and display the results of the queries. The accessing step requires that the following tasks must be performed:

- Set up an intermediate layer for the front-end tool to use. This layer, the metalayer, translates database structures and object names into business terms, so that the end user can interact with the data mart using terms that relate to the business function.
- Maintain and manage these business interfaces.
- Set up and manage database structures, like summarized tables, that help queries submitted through the front-end tool execute quickly and efficiently.

5. Managing

This step involves managing the data mart over its lifetime. In this step, management tasks are performed such as :

- Providing secure access to the data.
- Managing the growth of the data.
- Optimizing the system for better performance.
- Ensuring the availability of data even with system failures.

Summary

- Operational Systems can be defined as OLTP system that are used to a business. They run mission critical applications. RDBMS used for OLTP.
- **Examples of Operational Data:**

Data	Industry	Usage	Technology	Volumes
Customer File	All	Track Customer Details	Legacy application, flat files, main frames	Small-medium
Account Balance	Finance	Control account activities	Legacy applications, hierarchical databases, mainframe	Large
Point-of-Sale Data	Retail	Generate bills, manage stock	ERP, Client/Server, Relational databases	Very Large
Call Record	Telecommunications	Billing	Legacy application, hierarchical database, mainframe	Very Large
Production	Manufacturing	Control Production	ERP, Relational Databases, AS/400	Medium

■ OLTP vs Data Warehouse

	OLTP	Data Warehouse
1.	Application Oriented	Subject Oriented
2.	Used to run business	Used to analyze business
3.	Detailed data	Summarized and refined
4.	Current up to date	Snapshot data
5.	Isolated data	Integrated data
6.	Repetitive access	Ad-hoc access
7.	Clerical User	Knowledge User (Manager)
8.	Performance sensitivie	Performance relaxed
9.	Few Records accessed at a time (tens)	Large volumes accessed at a time (millions)
10.	Read/Update Access	Mostly Read (Batch Update)
11.	No data redundancy	Redundancy present
12.	Database Size 100 MB-100 GB	Database Size 100 GB - few terabytes

■ Meta Data is needed to

1. Share resources like users, tools.
2. Document system.

■ A Coarse Data Warehouse is often chosen when the organization has a relatively clean/new operational system and management wants to make the operational data more easily available for just that system.

■ A Central Data Warehouse; is often chosen when the organization has a clear understanding about it Information Access needs and wants to provide "quality", "integrated" , information to its knowledge workers.

■ A Distributed Data Warehouse is similar in most respects to a Central Data Warehouse, except that the data is distributed to separate mini-Data Warehouses (Data Marts)on local or specialized servers.

■ Difference between DW and DMarts **:** The following table summarizes the basic differences between a data warehouse and a data mart:

	Data Warehouse	Data Mart
Scope	Corporate	Line-of-Business (LoB)
Subjects	Multiple	Single Subject
Data Sources	Many	Few
Size (typical)	100 GB-TB+	< 100GB
Implementation Time	Months to years	

■ Difference between dependent and independent datamarts. The main difference between independent and dependent data marts is how the data mart is populated; that is, how you get data out of the sources and into the data mart. This step, called the Extraction-Transformation-and Loading (ETL) process, involves moving data from operational systems, filtering it, and loading it into the data mart.

With dependent data marts, this process is somewhat simplified because formatted and summarized (clean) data has already been loaded into the central data warehouse. The ETL process for dependent data marts is mostly a process of identifying the right subset of data relevant to the chosen data mart subject and moving a copy of it, perhaps in a summarized form. With independent data marts, however, you must deal with all aspects of the ETL process, much as you do with a central data warehouse. The number of sources is likely to be fewer and the amount of data associated with the data mart is less than the warehouse, given your focus on a single subject.

The motivations behind the creation of these two types of data marts are also typically different. Dependent data marts are usually built to achieve improved performance and availability, better control, and lower telecommunication costs resulting from local access of data relevant to a specific department. The creation of independent data marts is often driven by the need to have a solution within a shorter time.

Review Questions

1. Explain 3 tier architecture of data warehouse.
2. Compare 2 tier, 3 tier architecture and write the advantages and disadvantages of one over another.
3. Is there any difference between Datamarts and data warehouse, if yes, explain.
4. Why staging area is needed in data warehouse architecture ?
5. Explain the advantages and disadvantages of data granularity in data warehouse.
6. Discuss the issues while designing data warehouse.
7. Explain ETL process.
8. Discuss the characteristics of DW.
9. What are different components of DW ?
10. Discuss the importance of METADATA in DW.

❑❑❑

3

Chapter

Dimensional Modeling

INSIDE THIS CHAPTER

3.1. Introduction **3.2.** Dimensional Model vs ER Model **3.3.** Data Warehouse Schema Architecture **3.4.** A Data Mining Query Language, DMQL: Language Primitives **3.5.** Aggregate Tables **3.6.** Data Cubes **3.7.** Query Processing **3.8.** Strengths of Dimensional Model **3.9.** ER vs DM

This chapter covers the concept of dimensional modeling. Different schemas that can be used to organize the facts and tables will be discussed in this chapter. After studying this chapter students will be able to differentiate between the data warehouse schema architectures.

3.1. INTRODUCTION

Dimensional data model is most often used in data warehousing systems. Data model is a logical design technique that seeks to present the data in a standard, intuitive framework that allows for high-performance access. It is a different way to view the data in the database. This view may be used in Decision Support Systems (DSS) in addition with data mining tasks. Dimensional Modeling does not necessarily involve a relational database. But it can be implemented using a relational or a multidimensional DBMS.

A dimension is a collection of logically related attributes and is viewed as an axis for modelling the data. For example, the time dimension could be divided into many granularities like year which can be divided into quarters, quarters consists of months and weeks that consists of days (Fig. 3.1)

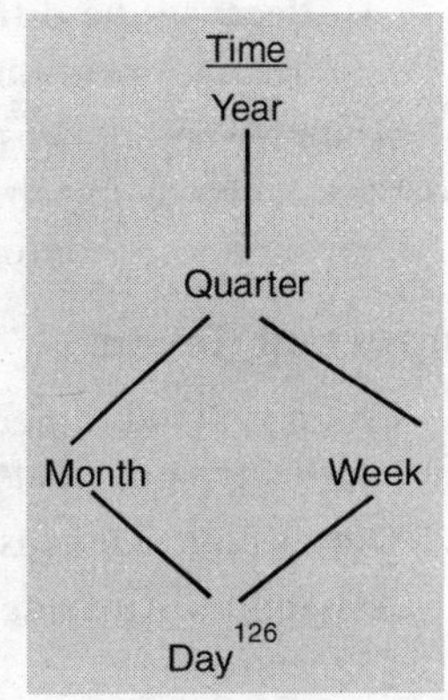

Fig. 3.1.

The specific data stored are called facts and usually are the numeric data. Thus Dimensional modeling always uses the concepts of facts (measures), and dimensions (context). DSS queries may access the facts from nay different dimensions and levels.

Facts and Dimensions: Facts are typically (but not always) numeric values that can be aggregated, and dimensions are groups of hierarchies and descriptors that define the facts.

Every dimensional model is composed of one table with a multipart key, called the fact table, and a set of smaller tables called dimension tables.

Each dimension table has a single-part primary key that corresponds exactly to one of the components of the multipart key in the fact table.

For example, sales amount is a fact; timestamp, product, register#, store#, etc. are elements of dimensions.

Dimensional models are built by business process area, *e.g.*, store sales, inventory, claims, etc. Because the different business process areas share some but not all dimensions, efficiency in design, operation, and consistency, is achieved using conformed dimensions, *i.e.*, using one copy of the shared dimension across subject areas. The term "conformed dimensions" was originated by kamball

3.2. DIMENSIONAL MODEL VS ER MODEL

As from the above discussion it is clear that dimensional modeling is used for designing the tables for a data warehouse, ER model is used for designing the tables for the operational systems. There are some differences between the two, described below:

1. The key to understanding the relationship between Dimensional Modeling (DM) and ER is that a single ER diagram breaks down into multiple DM diagrams, or 'stars'.
2. In dimensional modeling the denormalized tables are kept but in ER modeling the main aim is to remove the redundancy by normalizing the tables.
3. A dimensional model is designed to answer the queries on the overall business process but an ER model is well suited to answer the queries at the transaction level.

3.2.1. Need for a DM (Dimensional Model)

A dimensional model is created to provide the following information to the data warehouse developers.

1. The level of detail of data (granularity).
2. The facts that are to be included in the logical structure.
3. The process which includes the set of all the subject areas that are actually the logical structures which have to be designed.
4. Because the different business process areas share some but not all dimensions, efficiency in design, operation, and consistency, is achieved using conformed dimensions

A conformed dimension is a set of data attributes that have been physically implemented in multiple database tables using the same structure, attributes, domain values, definitions and concepts in each implementation. A conformed dimension cuts across many facts.

Important Terms

To understand dimensional data modeling and schemas, let's define some of the terms commonly used in this type of modeling:

Dimension: Dimension is a category of information. For example, the time dimension.

Attribute: Attribute is a unique level within a dimension. For example, Month is an attribute in the Time Dimension.

Hierarchy: The specification of levels that represents relationship between different attributes within a dimension. For example, one possible hierarchy in the Time dimension is Year ? Quarter ? Month ? Day.

Fact Table: A fact table is a table that contains the measures of interest. A fact table typically has two types of columns: foreign keys to dimension tables and measures those that contain numeric facts.

A fact table can contain fact's data on detail or aggregated level. For example, sales amount would be such a measure. This measure is stored in the fact table with the appropriate granularity. For example, it can be sales amount by store by day. In this case, the fact table would contain three columns: A date column, a store column, and a sales amount column.

Lookup Table: The lookup table provides the detailed information about the attributes. For example, the lookup table for the Quarter attribute would include a list of all of the quarters available in the data warehouse. Each row (each quarter) may have several fields, one for the unique ID that identifies the quarter, and one or more additional fields that specifies how that particular quarter is represented on a report (for example, first quarter of 2001 may be represented as "Q1 2001" or "2001 Q1").

A dimensional model includes fact tables and lookup tables. Fact tables connect to one or more lookup tables, but fact tables do not have direct relationships to one another. Dimensions and hierarchies are represented by lookup tables. Attributes are the non-key columns in the lookup tables.

In designing data models for data warehouses/data marts, the most commonly used schema types are Star Schema and Snowflake Schema.

3.3. DATA WAREHOUSE SCHEMA ARCHITECTURE [MDU Dec. 2011; May 2012]

There are many schema models designed for data warehousing but the most commonly used are:

- Star schema
- Snowflake schema
- Fact constellation schema.

The determination of which schema model should be used for a data warehouse should be based upon the analysis of project requirements, accessible tools and project team preferences.

3.3.1. Star Schema

The star schema architecture is the simplest data warehouse schema. It is called a star schema because the diagram resembles a star, with points radiating from a center. The center of the star consists of fact table and the points of the star are the dimension tables. Usually the fact tables in a star schema are in third normal form (3NF) whereas dimensional tables are de-normalized. Despite the fact that the star schema is the simplest architecture, it is most commonly used nowadays and is recommended by Oracle.

In the star schema design, a single object (the fact table) sits in the middle and is radially connected to other surrounding objects (dimension lookup tables) like a star. Each dimension is represented as a single table. The primary key in each dimension table is related to a foreign key in the fact table example: Assume our data warehouse keeps store sales data, and the different dimensions are time, store, product, and customer. The lines between two tables indicate that there is a primary key / foreign key relationship between the two tables. Note that different dimensions are not related to one another.

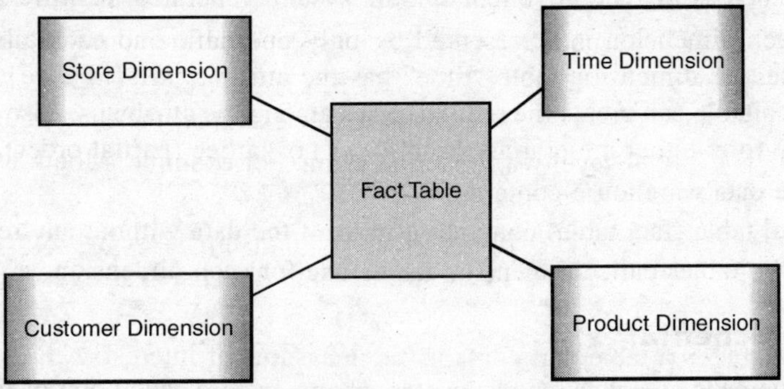

Fig. 3.2. Star Schema.

The main characteristics of star schema are:

- Simple structure: It is easy to understand star schema because star schema has a simple structure.
- Great query effectives: In this schema there are small number of tables, that are joined, thus respond effectively.
- Relatively long time of loading data into dimension tables: Size of the table could be large because of de-normalization, redundancy data.
- The star schema is most commonly used in the data warehouse implementations: This schema is widely supported by a large number of business intelligence tools.

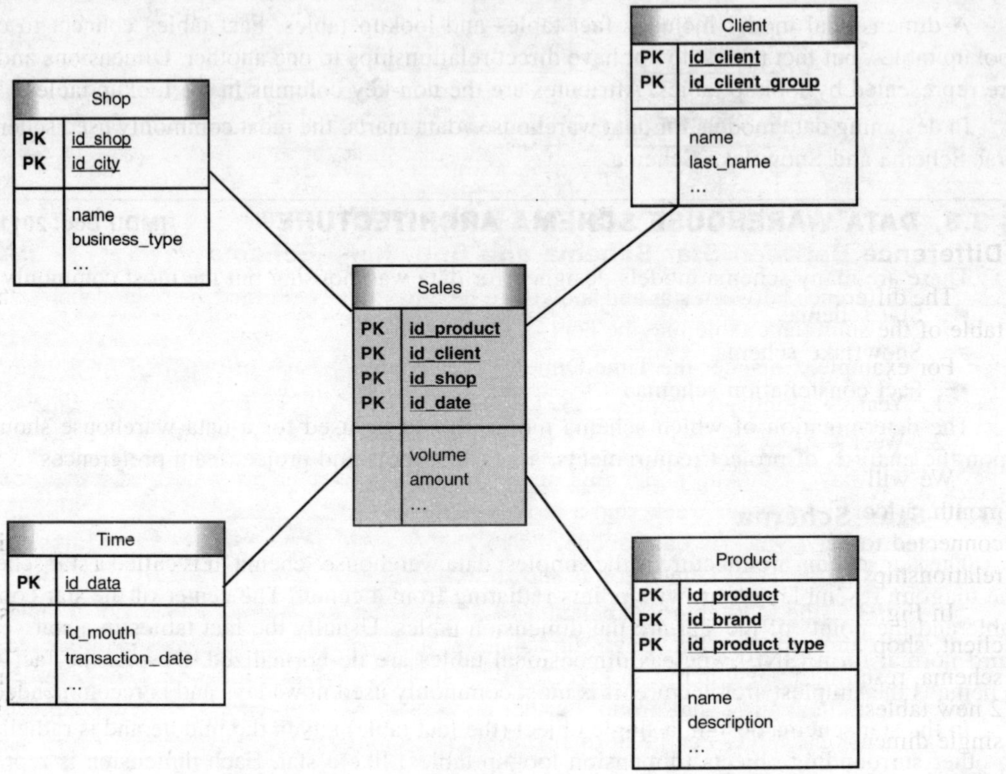

Fig. 3.3. Star Schema Example.

The schema contains a central fact table "sales" as shown in Fig. 3.3 that contains the keys to each of the 4 dimensions along with 2 measures volume and amount. To minimize the size of the fact table the dimension identifiers such as id_date, id_client etc. are system generated identifiers.

In this schema each dimension is represented by only one table and each table contains a set of attributes. For example, the dimension table "time" has the attribute sets{id_date}. But sometimes the entries in the dimension table can create the redundancies among the attributes. Also the attributes within a dimension table may form either a hierarchy (total order) or lattice (partial order).

In this schema the data warehouse contains;

1. A large central table (fact table) containing most of the data without any redundancies.
2. A set of smaller tables called dimension tables one for each dimension.

3.3.2. Snowflake Schema

The snowflake schema is an extension of the star schema, where each point of the star explodes into more points. In a star schema, each dimension is represented by a single dimensional table, whereas in a snowflake schema, that dimensional table is normalized into multiple lookup tables, each representing a level in the dimensional hierarchy. The resulting schema forms a shapc similar to a snowflake.

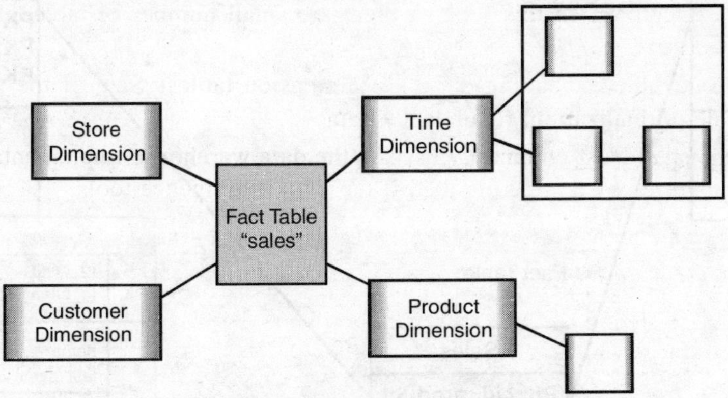

Fig. 3.4. Sample Snowlake Schema.

Difference Between Star Schema and Snowflake Schema **[MDU Dec. 2011]**

The difference between star and snowflake schema is that to reduce the redundancies, the dimensions table of the snowflake table may be kept in normalized form.

For example, Consider the Time Dimension that may consists of two different hierarchies:

1. Year → Month → Day
2. Week → Day

We will have 4 lookup tables in a snowflake schema: A lookup table for year, a lookup table for month, a lookup table for week, and a lookup table for day. Year is connected to Month, which is then connected to Day. Week is only connected to Day. A sample snowflake schema illustrating the above relationships in the Time Dimension is shown to the right in square (Fig. 3.4)

In Fig. 3.5, the fact table is labelled as "sales". There are 4 single dimensional tables named time, client, shop and product. The single dimensional table in the star schema is normalized in snowflake schema, resulting in new items. In Fig. 3.5, the single dimensional table for client can be normalized to 2 new tables: client and client group. Further normalization can be done on group name. Similarly other single dimensional tables can be normalized to give birth to snowflake schema.

Advantage of Snowflake Schema

1. Since in this schema the dimension table is kept in normalized form thus redundancies are reduced.
2. Other advantage of the snowflake schema is the improvement in query performance due to minimized disk storage requirements and joining smaller lookup tables.

Disadvantage

1. The main disadvantage of the snowflake schema is the additional maintenance efforts needed due to the increase number of lookup tables.
2. The snowflake schema can reduce the effectiveness of browsing since more joins will be needed to execute the query.

NOTE

Snowflake schema is not as popular as star schema in data warehouse design as the snowflake schema architecture is a more complex variation of the star schema used in a data warehouse, because the tables which describe the dimensions are normalized.

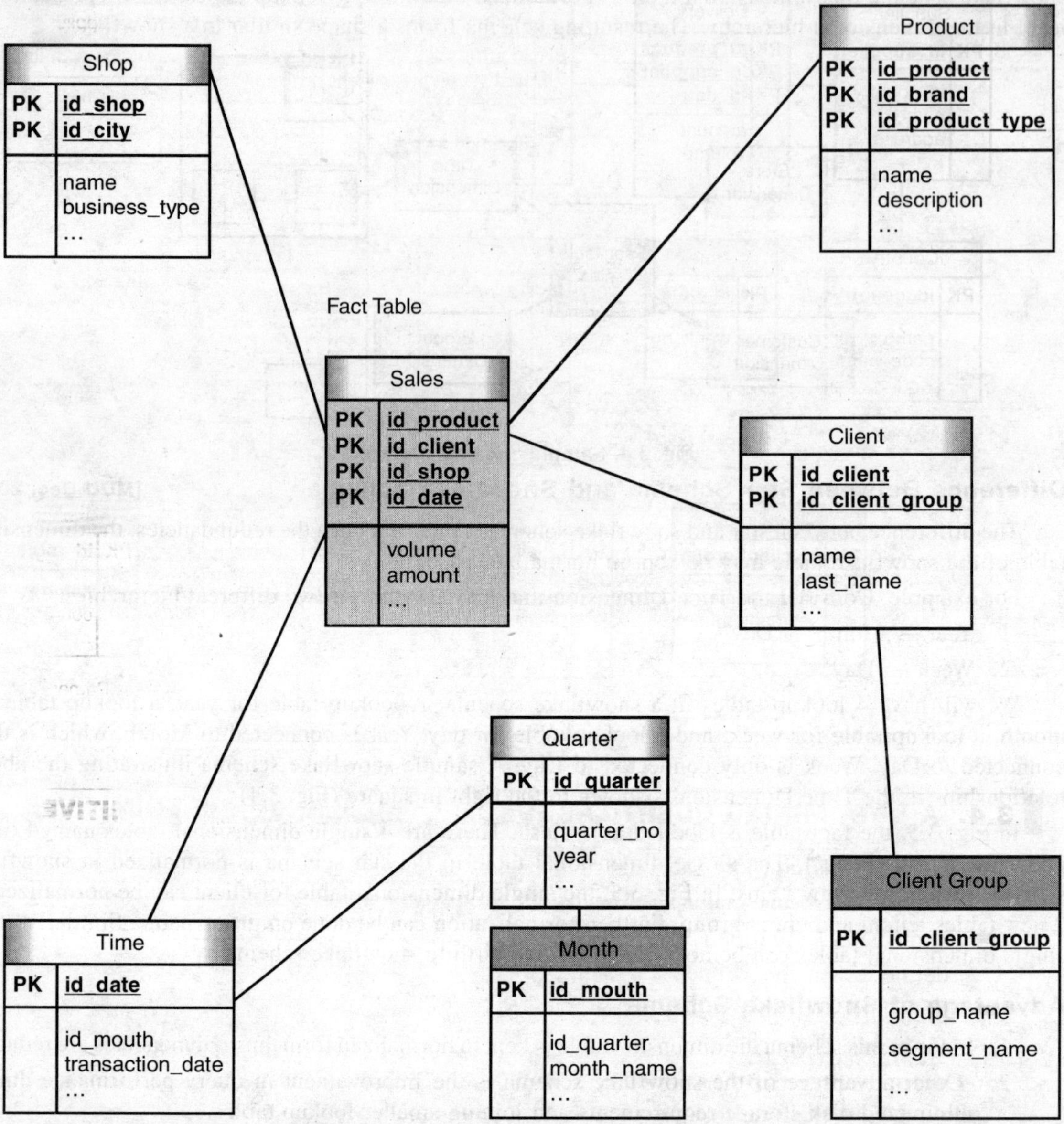

Fig. 3.5. Fact Tables and Dimensional Table.

3.3.3. Fact Constellation Schema/Glaxy Schema

For each star schema it is possible to construct fact constellation schema. It can be done by splitting the original star schema into more star schemes each of them describes facts on another level of dimension hierarchies. The fact constellation architecture contains multiple fact tables that share many dimension tables.

The main shortcoming of the fact constellation schema is a more complicated design because many variants for particular kinds of aggregation must be considered and selected. Moreover, dimension tables are still large.

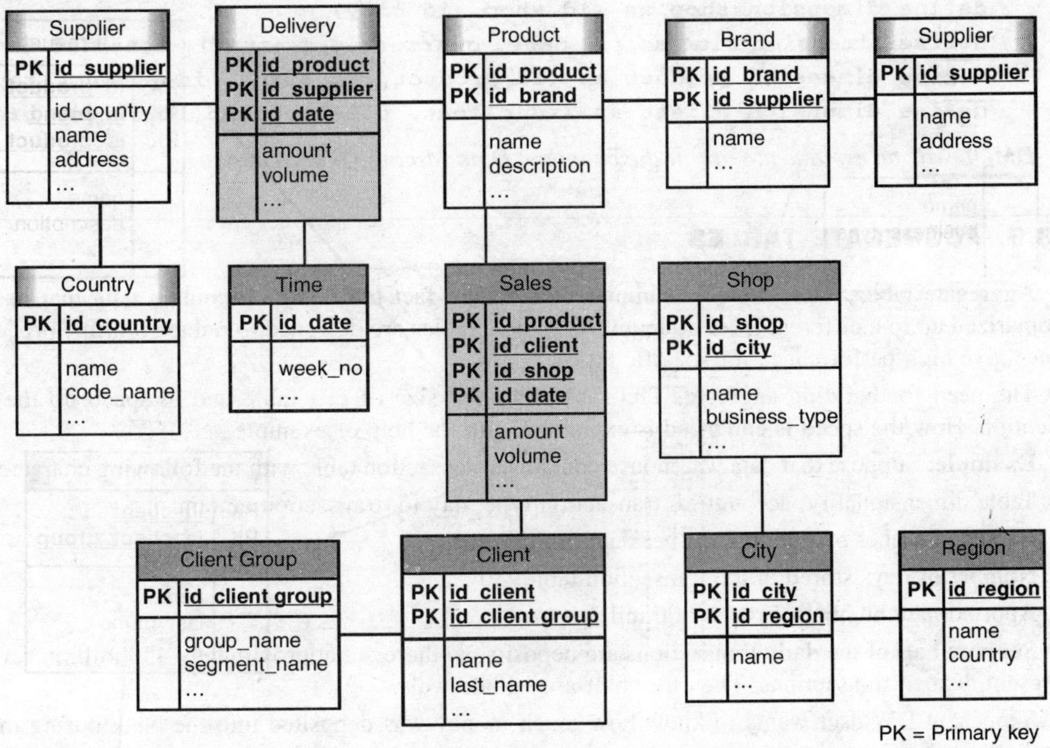

Fig. 3.6. Fast Constellation Schema.

3.4. A DATA MINING QUERY LANGUAGE, DMQL: LANGUAGE PRIMITIVES

➡ Cube Definition (Fact Table)

define cube <cube_name> [<dimension_list>]: <measure_list>

➡ Dimension Definition (Dimension Table)

define dimension <dimension_name> as (<attribute_or_subdimension_list>)

➡ Special Case (Shared Dimension Tables)

First time as "cube definition"

define dimension <dimension_name> as <dimension_name_first_time> in cube <cube_name_first_time>

Defining a Star Schema in DMQL

define cube sales_star [time, shop,client, product]:

```
amount = sum(sales),  volume = count(*)
define dimension time as (id_date)
define dimension shop as (id_shop, id_city)
define dimension product as (id_product, id_brand, id_product key)
define dimension client as (id_client, id_client group).
```

Defining a Snowflake Schema in DMQL

```
define cube sales_snowflake [time, shop,client, product]:
amount = sum(sales),  volume = count(*)
```

```
define dimension shop as (id shop, id city)
define dimension time as (id date, quarter(id quarter), month(id month))
define dimension product as (id product, id brand, id product key)
define dimension client as (id client, client group(id client group))
```

DMQL will be explained more in the chapter "Data Mining Query Language".

3.5. AGGREGATE TABLES

Aggregate tables, also know as summary tables, are fact tables which contain data that has been summarized up to a different level of detail. Aggregate tables are designed to reduce runtime processing. Hence give high performance for specific tasks.

The need for building aggregate fact tables is large size of fact table and to speed up the query execution. How the speed is enhanced is explained with the help of example.

Example: Suppose that data warehouse contains a transaction table with the following characteristics

Table dimensionality: account id, transaction type, day id, transaction amount

Average number of transactions per day: 30 million

Number of days stored in the transaction table: 30

Approximate number of rows: 900 million rows

Suppose half of the daily transactions are deposits, so there are approximately 450 million rows that represent deposit transactions. The other half are withdrawals.

Suppose a DW user wants to know how much money was deposited into the bank during the past month, the query is select sum(transaction_amount) from transaction_fact where transaction_type='deposit'

DW platform can scan 10 million rows per second; therefore, the approximate time to complete the query will be:

```
query time = number of rows / scan rate
```

which in our example translates into:

```
query time = 900 million rows / 10 million rows/second
query time = 90 seconds
```

Waiting 90 seconds for such a simple query is simply unacceptable, so here is where an aggregate table can help you. To answer our hypothetical question, we will build an aggregate table which summarizes the transaction table by transaction type. The aggregate may be defined as follows:

```
create table fact_transaction_aggregate as
select day_id, transaction_type, sum(transaction_amount) as
transaction_amount
from transaction_fact
group by day_id, transaction_type
```

We said before that there are only two transaction types and thirty days of data. Using the simplifying assumption that half of the daily transactions are deposits and half are withdrawals, the size of the new table will be only 60 rows! (table size = 30 days * 2 transaction types)

The SQL needed to get the answer is:

```
select sum(transaction_amount)
from transaction_aggregate
where transaction_type='deposit'
```

And the answer will come back in 0.000006 seconds (60 rows / 10 million rows /second).

Thus response is better using aggregate fact tables.

```
Fact tables contains million of records and retrieving the records
from fact table takes time. Where as aggregate table
contains limited data from all the required tables, and we
retrieve the data it takes less time.
```

3.6. DATA CUBES

[MDU May 2012; Dec. 2011]

Users of decision support systems often see data in the form of ***data cubes***. The cube is used to represent data along some measure of interest. Although called a "cube", it can be 2-dimensional, 3-dimensional, or higher-dimensional. Each dimension represents some attribute in the database and the cells in the data cube represent the measure of interest. For example, they could contain a count for the number of times that attribute combination occurs in the database, or the minimum, maximum, sum or average value of some attribute. Queries are performed on the cube to retrieve decision support information.

Example: We have a database that contains transaction information relating company sales of a part to a customer at a store location. The data cube formed from this database is a 3-dimensional representation, with each cell *(p,c,s)* of the cube representing a combination of values from *part, customer* and *store-location*. A sample data cube for this combination is shown in Figures 3.7 and 3.8. The contents of each cell is the count of the number of times that specific combination of values occurs together in the database. Cells that appear blank in fact have a value of zero. The cube can then be used to retrieve information within the database about, for example, which store should be given a certain part to sell in order to make the greatest sales.

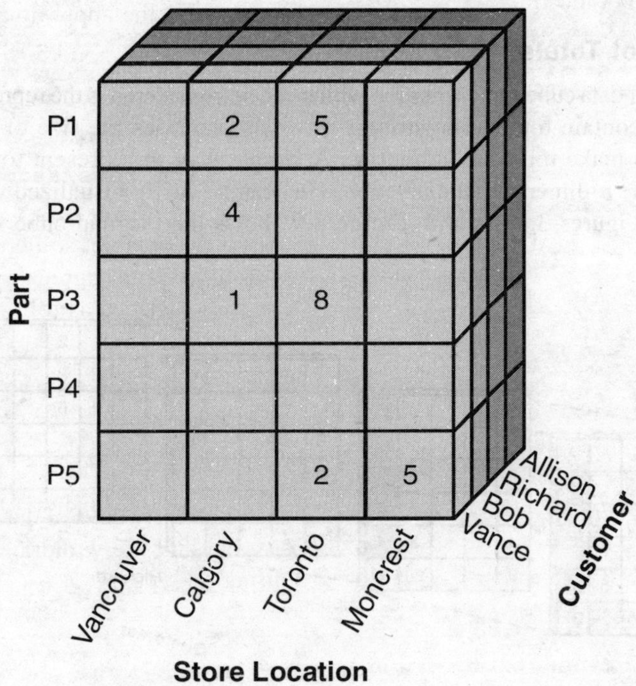

Fig. 3.7. Front View of Sample Data Cube.

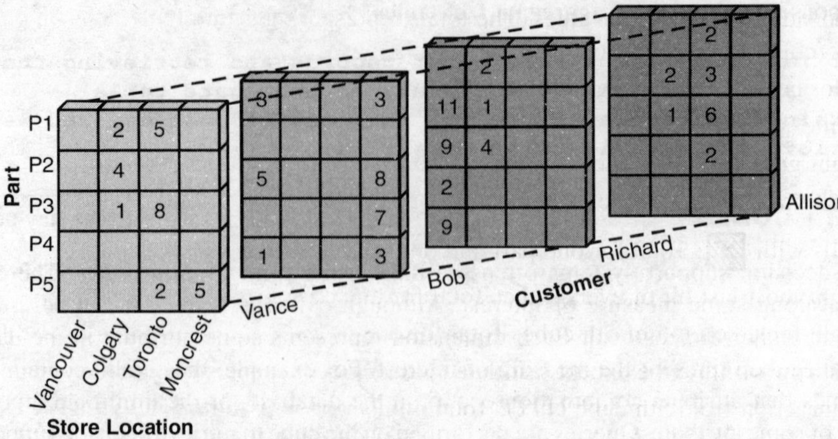

Fig. 3.8. Entire View of Sample Data Cube

3.6.1. Representation of Data Cubes

m-Dimensional Array

A data cube built from *m* attributes can be stored as an *m*-dimensional array. Each element of the array contains the measure value, such as count. The array itself can be represented as a 1-dimensional array. For example, a 2-dimensional array of size $x \times y$ can be stored as a 1-dimensional array of size $x*y$, where element (i, j) in the 2-D array is stored in location $(y*i+j)$ in the 1-D array. The disadvantage of storing the cube directly as an array is that most data cubes are sparse, so the array will contain many empty elements (zero values).

Representation of Totals

Another aspect of data cube representation which can be considered is the representation of totals. A simple data cube does not contain totals. The storage of totals increases the size of the data cube but can also decrease the time to make total-based queries. A simple way to represent totals is to add an additional layer on *n* sides of the *n*-dimensional data cube. This can be easily visualized with the 3-dimensional data cube introduced in Figures 3.7 and 3.8. Figure 3.9 shows the original cube with an additional layer on

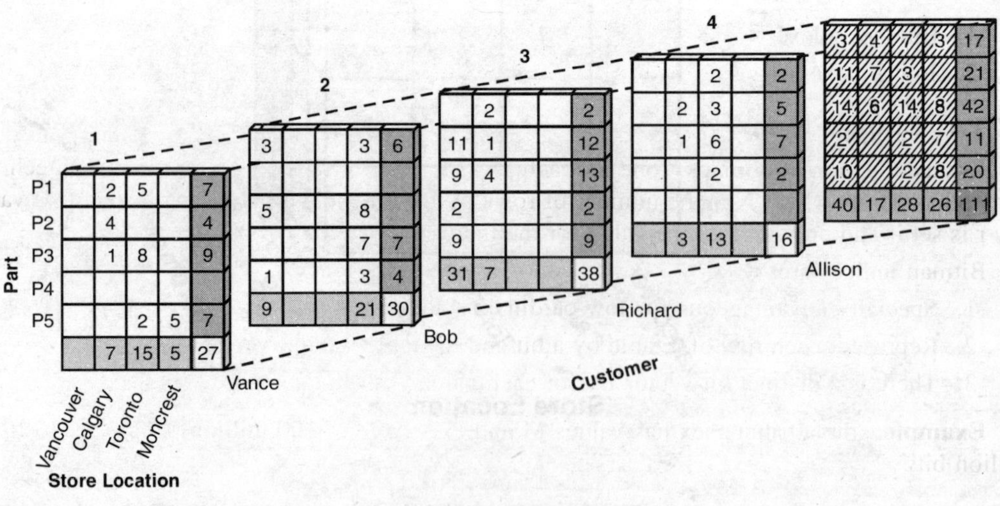

Fig. 3.9. Cube with Totals.

each of three sides to store total values. The totals represent the sum of all values in one horizontal row, vertical row (column) or depth row of the data cube.

The color coding used in Fig. 3.9 is as follows:

- White: Original values
- Light gray: Total for one customer and one store location
- Dark gray: Total for one customer and one part
- Cells with ▨: Total for one part and one store location
- Right most last bit of every cube: Total for one customer
- Right most column of 5th cube: Total for one part
- Last row of 5th cube except last data (111): Total for one store location
- Transaction from 5th cube (111): Total number of transactions in all

3.6.2. Operations on Data Cubes

Rollup or ***summarization***, drill down, slice and dice pivot, etc. are the operations done on data cubes and will be described in next chapter.

3.7. QUERY PROCESSING

Query processing includes:

1. Indexing
2. Aggregates
3. SQL extensions.

Recently, data warehouse system is becoming more and more important for decision-makers. Most of the queries against a large data warehouse are complex and iterative. The ability to answer these queries efficiently is a critical issue in the data warehouse environment. With multidimensional data the indices help to reduce the overhead of scanning the extremely large tables.

If the right index structures are built on columns, the performance of queries, especially ad hoc queries will be greatly enhanced. Types of indexing techniques are:

3.7.1. Indexing

1. Bitmap Indexes
2. Join Indexes

3.7.1.1. Bitmap index

It is a collection of bitmaps—one for each distinct value of the column. In this index technique, each bitmap has N bits where N is the number of rows in the table and a bit corresponding to a value v for a row r is set if and only if r has the value for the indexed attribute

Bitmap indices are:

1. Specially advantageous for low-cardinality domains
2. Represent each row of a table by a bit and the table as a bit vector
3. There is a distinct bit vector Bv for each value v for the domain

Example: the attribute sex has values M and F. A table of 100 million people needs 2 lists of 100 million bits

Gender	vote	Gender(M)	vote
M	Y	1	1
F	Y	0	1
F	N	0	0
M	N	1	0
F	Y	0	1
F	N	0	0

SQL query for the above can be written as:

```
select * from customer where
gender = 'M' and vote = 'Y'
```

The resultant is

```
1
0
0
0
0
0
```

Bitmap indices facilitate easy functions such as join and aggregation through the use of bitmap operations. There is significant reduction in space as compared to more traditional indices where the pointers to the records have to be maintained. The products that support bitmaps are: Model204, oracle 7.3

3.7.1.2. Join Index

Join indices support precomputed joins. When used for multidimensional data a common approach is to create a join index between the dimensional table and a fact table. A join index between a fact table and a dimension table correlates a dimension tuple with the fact tuples that have the same value on the common dimensional attribute. Join indexes can be created using bitmaps as opposed to pointers.

For example, a join index on *city* dimension of *calls* fact table

correlates for each city the calls (in the *calls* table) from that city.

Join indexes can also span multiple dimension tables

For example, a join index on *city* and *time* dimension of *calls* fact table

Star Join Processing

Use join indices to join the fact and dimension table

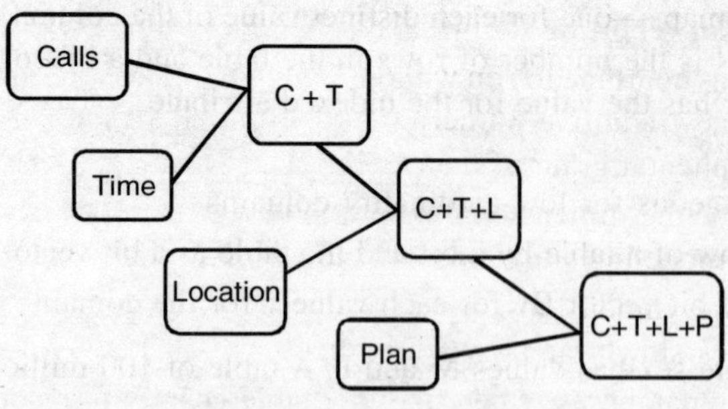

Fig. 3.10.

Optimization

The optimization can be achieved using select queries.

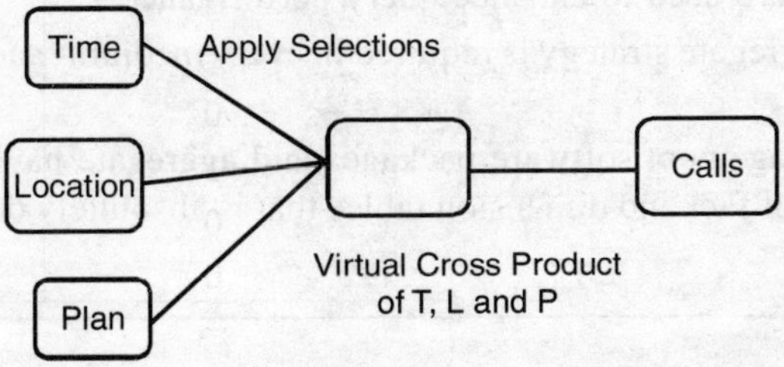

Fig. 3.11.

To speed up the query processing, the join indexing and the bitmap indexing methods can be integrated to form bitmapped join indices.

3.8. STRENGTHS OF DIMENSIONAL MODEL

1. The dimensional model has a number of important data warehouse advantages that the ER model lacks.
   - First, the dimensional model is a predictable, standard framework. Report writers, query tools, and user interfaces can all make strong assumptions about the dimensional model to make the user interfaces more understandable and to make processing more efficient.
   - Rather than using a cost-based optimizer, a database engine can make very strong assumptions about first constraining the dimension tables and then "attacking" the fact table all at once with the Cartesian product of those dimension table keys satisfying the user's constraints.
2. A second strength of the dimensional model is that the predictable framework of the star join schema withstands unexpected changes in user behaviour. Every dimension is equivalent. All dimensions can be thought of as symmetrically equal entry points into the fact table. The logical design can be done **independent of expected query patterns**. The user interfaces are symmetrical, the query strategies are symmetrical, and the SQL generated against the dimensional model is symmetrical.
3. A third strength of the dimensional model is that it is 'gracefully extensible' to accommodate unexpected new data elements and new design decisions.

   Gracefully extensible means that:
   - all existing tables (both fact and dimension) can be changed in place by simply adding new data rows in the table, or the table can be changed in place with a SQL alter table command.
   - Data should not have to be reloaded.
   - No query tool or reporting tool needs to be reprogrammed to accommodate the change.
   - Old applications continue to run without yielding different results. Adding new unanticipated facts (that is, new additive numeric fields in the fact table), as long as they are consistent with the fundamental grain of the existing fact table
4. A fourth strength of the dimensional model is that there is a body of standard approaches for handling common modeling situations in the business world. These modeling situations include:
   - Slowly changing dimensions, where a "constant" dimension such as Product or Customer actually evolves slowly and asynchronously.
   - Event-handling databases, where the fact table usually turns out to be "factless."
   - Many others

5. A final strength of the dimensional model is the management of aggregates.

   Aggregates are summary records that are logically redundant with base data already in the data warehouse, but they are used to enhance query performance.

   A comprehensive aggregate strategy is required in every medium- and large-sized data warehouse implementation.

All of the aggregate management software packages and aggregate navigation utilities depend on a very specific single structure of fact and dimension tables that is absolutely dependent on the dimensional model.

3.9. ER VS DW

ER models are not appropriate for Data Warehouses. ER modeling does not really model a business; rather, it models the micro relationships among data elements.

ER models are wildly variable in structure. As such, it is extremely difficult to optimize query performance.

Conversion from ER Model to Data Warehouse Architecture (Star Schema)

Think of a large ER diagram as representing every possible business process within an application. The ER diagram may have Sales Calls, Order Entries, Shipment Invoices, Customer Payments, and Product Returns, all on the same diagram. Thus the steps to convert ER model to DW are:

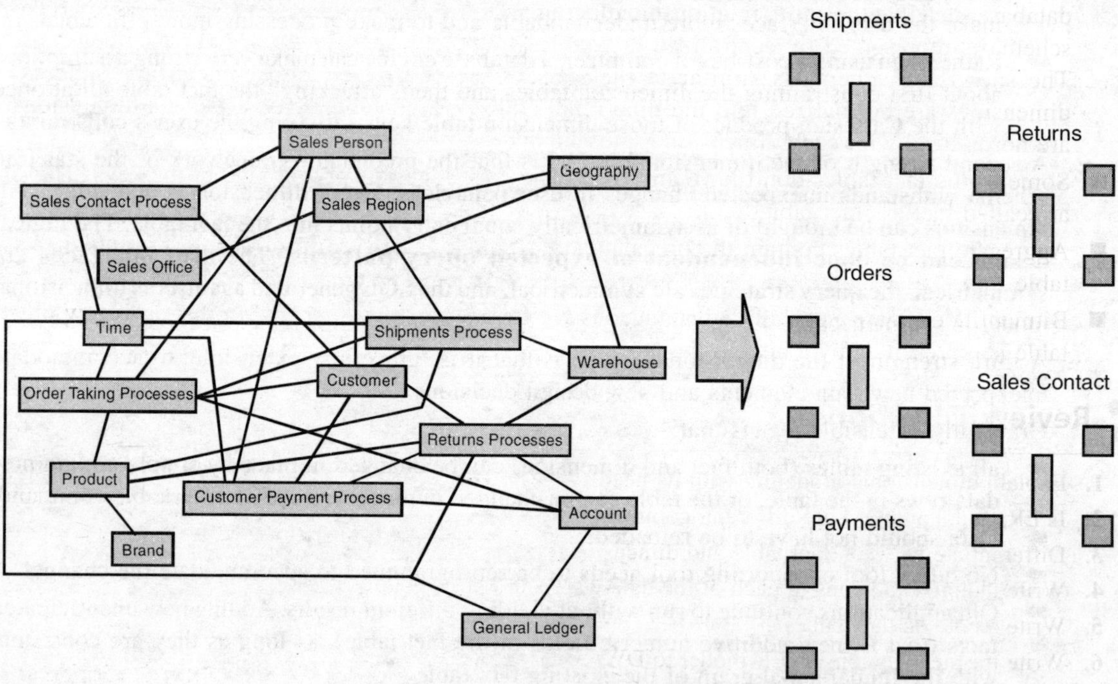

Fig. 3.12. Conversion from ER Model to Data Warehouse.

➡ To create the individual 'stars' that exist within an application:

➡ Look for many-to-many relationships in the ER model containing numeric and additive facts and to designate them as fact tables.

➡ Alternatively, look for 'events' or 'transactions' – these may also be facts.

➡ Denormalize all of the remaining tables into flat tables with single-part keys that connect directly to the fact tables. These tables become the dimension tables.

➡ In cases where a dimension table connects to more than one fact table, we represent this same dimension table in both schemas, and we refer to the dimension tables as "conformed" between the two dimensional models.

Summary

- Dimensional Modelling is a design concept used by many data warehouse designers to build their data warehouse. In this design model all the data is stored in two types of tables - Facts table and Dimension table. Fact table contains the facts/measurements of the business and the dimension table contains the context of measurements *i.e.,* the dimensions on which the facts are calculated.
- There are three types of data warehouse architecture schemas exists: star schema, snowflake schema and fact constellation.
- The **star schema** (also called **star-join schema**, **data cube**, or **multi-dimensional schema**) is the simplest style of data warehouse schema. The star schema consists of one or more fact tables referencing any number of dimension tables. The star schema is an important special case of the snowflake schema, and is more effective for handling simpler queries.
- In computing, a **snowflake schema** is a logical arrangement of tables in a multidimensional database such that the entity relationship diagram resembles a snowflake in shape. The snowflake schema is represented by centralized fact tables which are connected to multiple dimensions. The snowflake schema is similar to the star schema. However, in the snowflake schema, dimensions are normalized into multiple related tables, whereas the star schema's dimensions are normalized with each dimension represented by a single table.
- Some of the fact tables don't have any facts. They may consists of nothing but the keys. These are called factless fact tables.
- Aggregate fact tables contain the precalculated summaries derived from the most granular fact table.
- Bitmap indexes are Specially advantageous for low-cardinality domains. In this each row of a table is represented by a bit and the table as a bit vector

Review Questions

1. Explain dimensional modeling with its need.
2. Is ER model and DM are different, comment.
3. Differentiate between fact table and dimensional table.
4. Write the pros and cons of each of the data warehouse architecture scheme.
5. Write a note on data cubes.
6. Write the steps to convert ER model to DW.
7. Why dimensional model are used, explain with its strengths.
8. List down the features of fact table and a dimension table.
9. A fact table is narrow while dimension table is wide. Justify and explain with example.
10. Write a note on star schema with example, advantages and disadvantages.

4

Chapter

OLAP Strategies for Large Data Warehouse Projects

INSIDE THIS CHAPTER

4.1. Introduction **4.2.** The Need For Multidimensional Information **4.3.** OLAP and OLTP **4.4.** OLAP operations in Multidimensional Data Model **4.5.** OLAP Server/Models **4.6.** MOLAP—Multidimensional OLAP **4.7.** ROLAP—Relational OLAP **4.8.** Difference between ROLAP and MOLAP **4.9.** HOLAP—Hybrid OLAP **4.10.** DOLAP—Desktop OLAP **4.11.** OLAP Applications.

In this chapter on-line analytical processing will be discussed in detail. In this chapter we will learn how OLAP is used to perform multidimensional analysis. The functions, applications and different models of OLAP will be covered in this chapter. Also OLAP servers will be described in this chapter.

4.1. INTRODUCTION

Online analytical processing, or **OLAP** is an approach to answer multi-dimensional analytical queries OLAP is part of business intelligence

OLAP tools enable users to interactively analyze multidimensional data from multiple perspectives. OLAP consists of three basic analytical operations: consolidation (roll-up), drill-down, and slicing and dicing. Consolidation involves the aggregation of data that can be accumulated and computed in one or more dimensions. For example, all sales offices are rolled up to the sales department or sales division to anticipate sales trends. In contrast, the drill-down is a technique that allows users to navigate through the details. For instance, users can view the sales by individual products that make up a region's sales. Slicing and dicing is a feature whereby users can take out (slicing) a specific set of data of the cube and view (dicing) the slices from different viewpoints.

NOTE

OLAP Is FASMI: *Fast Analysis dimensional multidimensional information.*

4.2. THE NEED FOR MULTIDIMENSIONAL INFORMATION [MDU Dec. 2011]

Information, as it is usually stored in the present-day databases, is in the form of relations. Tables hold data, which can be queried on request and the information, retrieved. This is sufficient as long as one does not need to attach any semantics to this data. Once a person needs to acquire context and meaning of this information, and relationships to other pieces of information, simple querying is of little help. Information, here, must be viewed as slices or dimensions. Viewing information from different points implies that one can model different scenarios based on the same set of information, by just arranging them in different analytical formats.

Multidimensionality allows for assigning relationships between seemingly unrelated information fields. It allows combining information from different sources, and relating them to obtain useful knowledge.

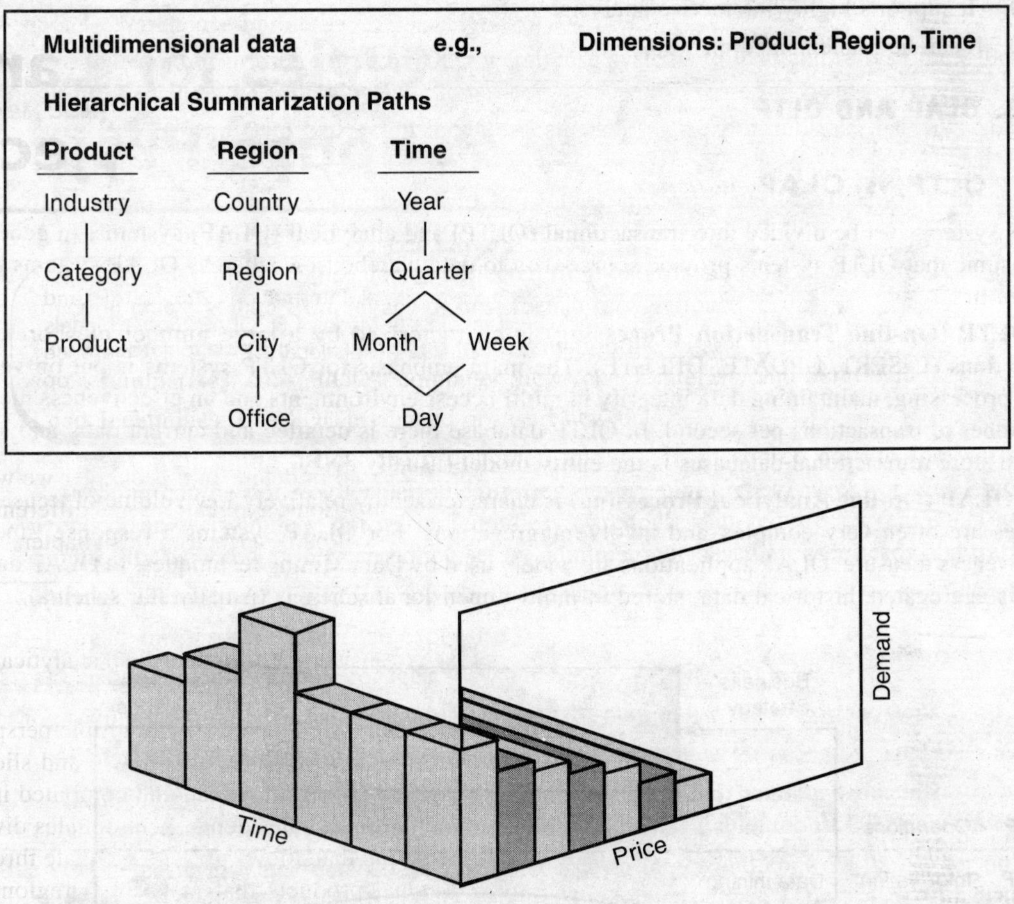

Fig. 4.1. Multidimensional Data.

Consider the example of the demand versus pricing data of a particular product. This data represents the sum total of all the demand in different regions of the business. It also aggregates demands over different time periods. However, for a better understanding of the real market demand, one needs to apply the dimensions of time, as well as the specific locations to the available data. This fact brings out the need for multidimensionality of information visualization and access.

Strengths of OLAP

1. It is a powerful visualization paradigm *i.e.*, it improves the presentation of results through visual presentation using graphs and charts
2. It provides fast, interactive response times
3. It is good for analyzing time series
4. It can be useful to find some clusters and outliers
5. Many vendors offer OLAP tools
6. It enables the analysts to measure the metrics along several dimensions.
7. It allow the data to be viewed from different perspective
8. It enables the application of complex mathematical formulae and calculations to measure for analysis.
9. It supports highly interactive analysis.
10. It can be implemented on web.

4.3. OLAP AND OLTP **[MDU May 2012]**

4.3.1. OLTP vs. OLAP

IT systems can be divided into transactional (OLTP) and analytical (OLAP) systems. In general we can assume that OLTP systems provide source data to data warehouses, whereas OLAP systems help to analyze it.

OLTP (On-line Transaction Processing) is characterized by a large number of short on-line transactions (INSERT, UPDATE, DELETE). The main emphasis for OLTP systems is put on very fast query processing, maintaining data integrity in multi-access environments and an effectiveness measured by number of transactions per second. In OLTP database there is detailed and current data, and schema used to store transactional databases is the entity model (usually 3NF).

- **OLAP (On-line Analytical Processing)** is characterized by relatively low volume of transactions. Queries are often very complex and involve aggregations. For OLAP systems a response time is an effectiveness measure. OLAP applications are widely used by Data Mining techniques. In OLAP database there is aggregated, historical data, stored in multi-dimensional schemas (usually star schema).

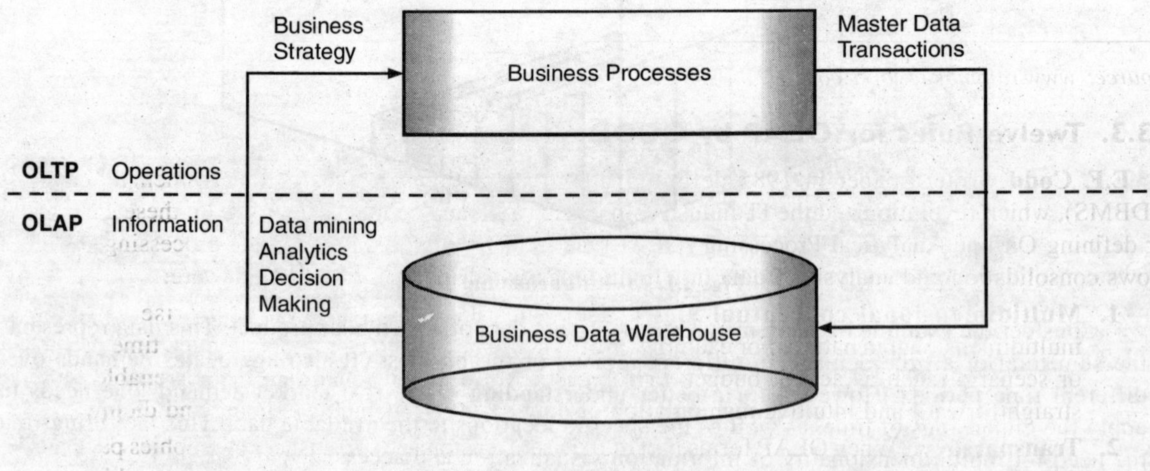

Fig. 4.2. OLAP and OLTP.

4.3.2. Differences Between OLTP and OLAP System

The following table summarizes the major differences between OLTP and OLAP system design.

	OLTP System Online Transaction Processing (Operational System)	**OLAP System Online Analytical Processing (Data Warehouse)**
Source of data	Operational data; OLTPs are the original source of the data.	Consolidation data; OLAP data comes from the various OLTP Databases
Purpose of data	To control and run fundamental business tasks	To help with planning, problem solving, and decision support
What the data	Reveals a snapshot of ongoing business processes	Multi-dimensional views of various kinds of business activities
Inserts and Updates	Short and fast inserts and updates initiated by end users	Periodic long-running batch jobs refresh the data
Queries	Relatively standardized and simple queries Returning relatively few records	Often complex queries involving aggregations
Processing Speed	Typically very fast	Depends on the amount of data involved; batch data refreshes and complex queries may take many hours; query speed can be improved by creating indexes
Space Requirements	Can be relatively small if historical data is archived	Larger due to the existence of aggregation structures and history data; requires more indexes than OLTP
Database Design	Highly normalized with many tables	Typically de-normalized with fewer tables; use of star and/or snowflake schemas
Backup and Recovery	Backup religiously; operational data is critical to run the business, data loss is likely to entail significant monetary loss and legal liability	Instead of regular backups, some environments may consider simply reloading the OLTP data as a recovery method

Source: www.rainmakerworks.com

4.3.3. Twelve Rules for OLAP by CODD

E.F. Codd wrote a paper in 1985 defining rules for Relational Database Management Systems (RDBMS), which revolutionized the IT industry. In 1993, Codd and colleagues worked up these 12 rules for defining On-line Analytical Processing (OLAP), an industry of software and data processing which allows consolidation and analysis of data in a multidimensional space. Codd's 12 rules are:

1. **Multidimensional conceptual view:** User-analysts would view an enterprise as being multidimensional in nature. For example, profits could be viewed by region, product, time period, or scenario (such as actual, budget, or forecast). Multi-dimensional data models enable more straightforward and intuitive manipulation of data by users, including "slicing and dicing".
2. **Transparency:** When OLAP forms part of the users' customary spreadsheet or graphics package, this should be transparent to the user. OLAP should be part of an open systems architecture

which can be embedded in any place desired by the user without adversely affecting the functionality of the host tool. The user should not be exposed to the source of the data supplied to the OLAP tool, which may be homogeneous or heterogeneous.

3. **Accessibility:** The OLAP tool should be capable of applying its own logical structure to access heterogeneous sources of data and perform any conversions necessary to present a coherent view to the user. The tool (and not the user) should be concerned with where the physical data comes from.
4. **Consistent reporting performance:** Performance of the OLAP tool should not suffer significantly as the number of dimensions is increased.
5. **Client/server architecture:** The server component of OLAP tools should be sufficiently intelligent that the various clients can be attached with minimum effort. The server should be capable of mapping and consolidating data between disparate databases.
6. **Generic dimensionality:** Every data dimension should be equivalent in its structure and operational capabilities.
7. **Dynamic sparse matrix handling:** The OLAP server's physical structure should have optimal sparse matrix handling.
8. **Multi-user support:** OLAP tools must provide concurrent retrieval and update access, integrity and security.
9. **Unrestricted cross-dimensional operations:** Computational facilities must allow calculation and data manipulation across any number of data dimensions, and must not restrict any relationship between data cells.
10. **Intuitive data manipulation:** Data manipulation inherent in the consolidation path, such as drilling down or zooming out, should be accomplished via direct action on the analytical model's cells, and not require use of a menu or multiple trips across the user interface.
11. **Flexible reporting:** Reporting facilities should present information in any way the user wants to view it.
12. **Unlimited dimensions and aggregation levels:** The number of data dimensions supported should, to all intents and purposes, be unlimited. Each generic dimensions should enable an essentially unlimited number of user-defined aggregation levels within any given consolidation path.

4.4. OLAP OPERATIONS IN MULTIDIMENSIONAL DATA MODEL

In Multiple dimension each dimension contains multiple levels of abstraction. This provides the users with the flexibility to view data from different perspectives. Some of the operations that a user can perform using OLAP are described below:

Slice: A slice is a subset of a multi-dimensional array corresponding to a single value for one or more members of the dimensions not in the subset.

Dice: The dice operation is a slice on more than two dimensions of a data cube (or more than two consecutive slices).

Drill Down/Up: Drilling down or up is a specific analytical technique whereby the user navigates among levels of data ranging from the most summarized (up) to the most detailed (down).

Roll-up: A roll-up involves computing all of the data relationships for one or more dimensions. To do this, a computational relationship or formula might be defined.

Pivot: To change the dimensional orientation of a report or page display.

4.5. OLAP SERVER/MODELS

OLAP is implemented in a multi-user client/server mode and offers consistently rapid response to queries, regardless of database size and complexity. OLAP helps the user to synthesize enterprise information through comparative, personalized viewing, as well as through analysis of historical and projected data in various "what-if" data model scenarios. This is achieved through use of an OLAP Server.

An OLAP server is a high-capacity, multi-user data manipulation engine specifically designed to support and operate on multi-dimensional data structures. A multi-dimensional structure is arranged so that every data item is located and accessed based on the intersection of the dimension members which define that item. The design of the server and the structure of the data are optimized for rapid ad-hoc information retrieval in any orientation, as well as for fast, flexible calculation and transformation of raw data based on formulaic relationships. The OLAP Server may either physically store the processed multi-dimensional information to deliver consistent and rapid response times to end users, or it may populate its data structures in real-time from relational or other databases, or offer a choice of both. Given the current state of technology and the end user requirement for consistent and rapid response times, staging the multi-dimensional data in the OLAP Server is often the preferred method.

4.6. MOLAP—MULTIDIMENSIONAL OLAP

Data is stored in the form of multidimensional cubes in a dedicated Multidimensional Database server over which the user can perform analysis. Data for known questions is aggregated and stored in a cube. Dimensional analysis is usually limited in MOLAP.

Advantages of MOLAP includes high performance.

Most MOLAP solutions store these data in an optimized multidimensional array storage, rather than in a relational database (*i.e.,* in ROLAP).

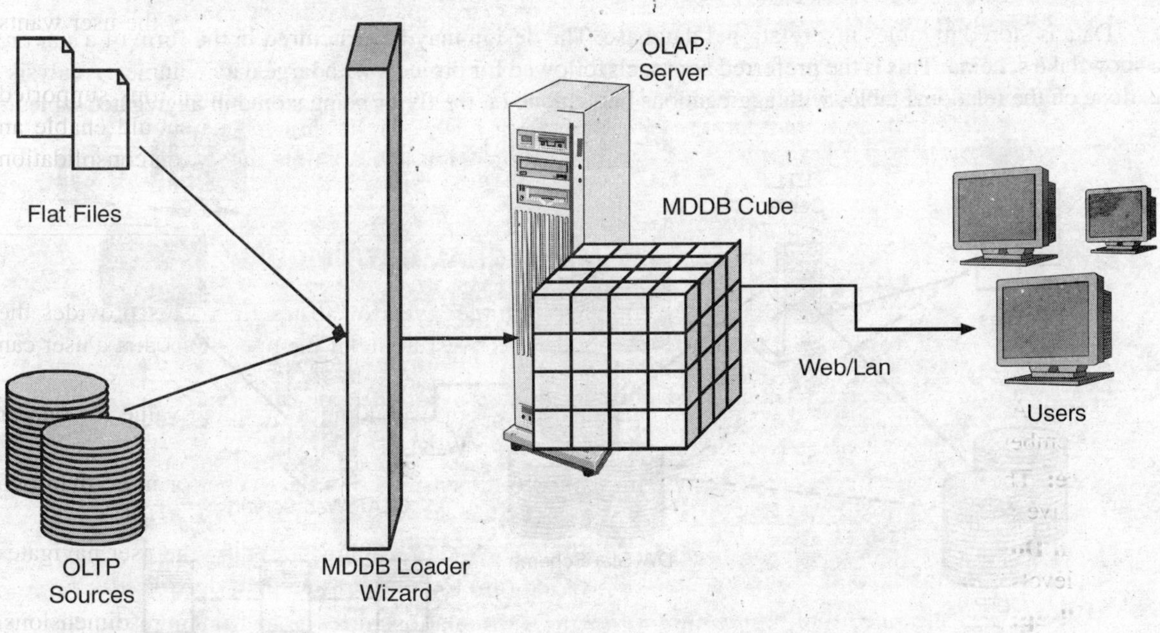

Fig. 4.3. MOLAP.

The advantages and disadvantages of MOLAP are :

Advantages of MOLAP

- MOLAP offers fast query performance due to optimized storage, multidimensional indexing and caching.
- On-disk size of data is small as compared to data stored in relational database due to compression techniques.
- MOLAP supports automated computation of higher level aggregates of the data.
- It is very compact for low dimension data sets.
- Array models provide natural indexing in MOLAP.
- Effective data extraction achieved through the pre-structuring of aggregated data.

Disadvantages of MOLAP

- Within some MOLAP Solutions the processing step (data load) can be quite lengthy, especially on large data volumes. This is usually remedied by doing only incremental processing, *i.e.,* processing only the data which have changed (usually new data) instead of reprocessing the entire data set.
- MOLAP tools traditionally have difficulty querying models with dimensions with very high cardinality (*i.e.,* millions of members).
- Some MOLAP products have difficulty updating and querying models with more than ten dimensions. This limit differs depending on the complexity and cardinality of the dimensions in question. It also depends on the number of facts or measures stored. Other MOLAP products can handle hundreds of dimensions.
- Some MOLAP methodologies introduce data redundancy.

4.7. ROLAP—RELATIONAL OLAP

Data is stored in tables in a relational database. The design may be structured in the form of a star or a snowflake schema. This is the preferred approach followed for projects with large data volumes. Analysis is done on the relational tables with aggregations being done on the fly or being stored in aggregated tables.

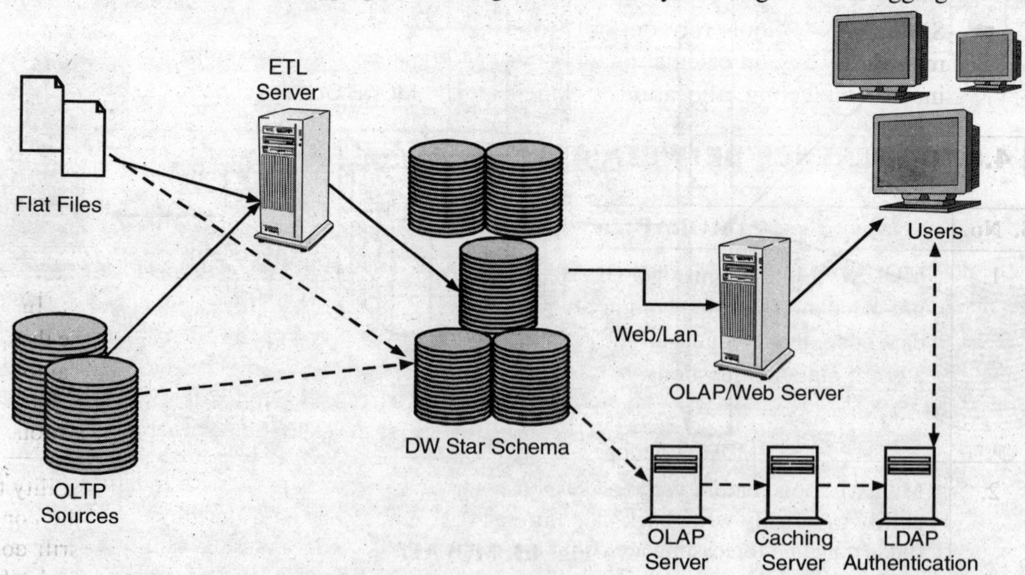

Fig. 4.4. Relational OLAP.

Advantages of ROLAP

- ROLAP is considered to be more scalable in handling large data volumes, especially models with dimensions with very high cardinality (*i.e.,* millions of members).
- With a variety of data loading tools available, and the ability to fine tune the ETL code to the particular data model, load times are generally much shorter than with the automated MOLAP loads.
- The data are stored in a standard relational database and can be accessed by any SQL reporting tool (the tool does not have to be an OLAP tool).
- ROLAP tools are better at handling *non-aggregatable facts* (*e.g.,* textual descriptions). MOLAP tools tend to suffer from slow performance when querying these elements.
- By decoupling the data storage from the multi-dimensional model, it is possible to successfully model data that would not otherwise fit into a strict dimensional model.
- The ROLAP approach can leverage database authorization controls such as row-level security, whereby the query results are filtered depending on preset criteria applied, for example, to a given user or group of users (SQL WHERE clause).

Disadvantages of ROLAP

- ROLAP tools have slower performance than MOLAP tools.
- The loading of *aggregate tables* must be managed by custom ETL code. The ROLAP tools do not help with this task. This means additional development time and more code to support.
- When the step of creating aggregate tables is skipped, the query performance then suffers because the larger detailed tables must be queried. This can be partially remedied by adding additional aggregate tables, however it is still not practical to create aggregate tables for all combinations of dimensions/attributes.
- ROLAP relies on the general purpose database for querying and caching, and therefore several special techniques employed by MOLAP tools are not available (such as special hierarchical indexing). However, modern ROLAP tools take advantage of latest improvements in SQL language such as CUBE and ROLLUP operators, DB2 Cube Views, as well as other SQL OLAP extensions. These SQL improvements can mitigate the benefits of the MOLAP tools.
- Since ROLAP tools rely on SQL for all of the computations, they are not suitable when the model is heavy on calculations which don't translate well into SQL. Examples of such models include budgeting, allocations, financial reporting and other scenarios.

4.8. DIFFERENCE BETWEEN ROLAP AND MOLAP **[MDU May 2012]**

S. No.	MOLAP	ROLAP
1.	MOLAP (Multidimensional OLAP) tools utilize a pre-calculated data set, commonly referred to as a data cube, that contains all the possible answers to a given range of questions.	ROLAP (relational OLAP) tools do not use pre-calculated data cubes. Instead, they intercept the query and pose the question to the standard relational database and its tables in order to bring back the data required to answer the question.
2.	MOLAP tools feature very fast response, and the ability to quickly write back data into the data set (budgeting and forecasting are common applications).	ROLAP tools feature the ability to ask any question (not limited to the contents of a cube) and the ability to drill down to the lowest level of detail in the database.

S. No.	MOLAP	ROLAP
3.	Primary downsides of MOLAP tools are limited scalability (the cubes get very big, very fast when you start to add dimensions and more detailed data), inability to contain detailed data (you are forced to use summary data unless your data set is very small), and load time of the cubes.	Primary downsides of ROLAP tools are slow response and some limitations on scalability (depending on the technology architecture that is utilized).
4.	The most common examples of MOLAP tools are Hyperion (Arbor) Essbase and Oracle (IRI) Express. MOLAP tools are best used for users who have "bounded" problem sets (they want to ask the same range questions every day/week/month on an updated cube, *e.g.*, finance).	The most common examples of ROLAP tools are MicroStrategy and Sterling (Information Advantage). ROLAP tools are best used for users who have "unbounded" problem set (they don't have any idea what they want to ask from day to day; *e.g.*, marketing). It is very important to pay close attention to the underlying architecture of ROLAP tools, as some tools are very "scalability challenged."

4.9. HOLAP—HYBRID OLAP

A mixture of MOLAP and ROLAP – HOLAP combines the strategies of storing massive granular data in an RDBMS and forming aggregated multidimensional cubes of data to facilitate quick summary reporting.

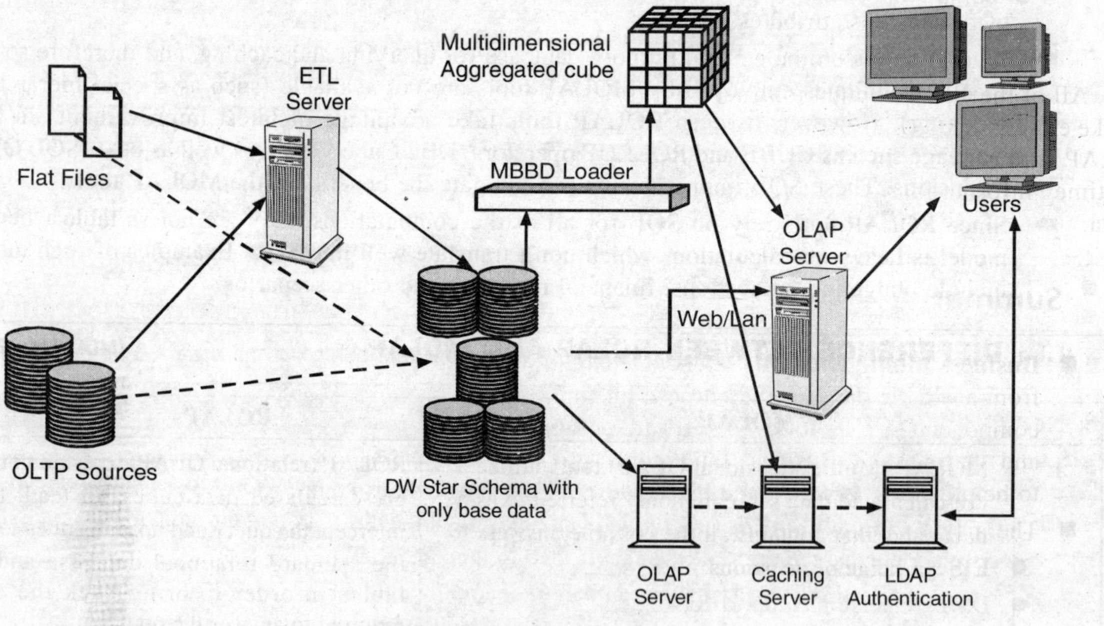

Fig. 4.5. Hybrid OLAP.

The OLAP server has the task of figuring out where to access the data from. In case of aggregated data requests the information is sought from the MDDB cube while granular queries are redirected to the relational database.

4.10. DOLAP—DESKTOP OLAP

Multidimensional micro cubes are formed and downloaded on the users desktop. Datasets is limited to the boundaries defined by the user with no access to granular data.

Based on the tool selection either or a combination of the above architectures can be implemented. It is best to start with ROLAP architecture for large voluminous data and move in to MOLAP cubes for aggregate executive reporting. (HOLAP)

4.11. OLAP APPLICATIONS

OLAP is widely used in several realms of data management. Some of these applications include:

1. **Financial Applications:**
   - Activity-based costing (resource allocation)
   - Budgeting
2. **Marketing/Sales Applications:**
   - Market Research Analysis
   - Sales Forecasting
   - Promotions Analysis
   - Customer Analyses
   - Market/Customer Segmentation
3. **Business modeling**
   - Simulating business behaviour
   - Extensive, real-time decision support system for managers

All of the above applications need ability to provide managers with the information they need to make effective decisions about an organization's strategic directions. The key indicator of a successful OLAP application is its ability to provide information, as needed, that is, its ability to provide "just-in-time" information for effective decision-making. This requires more than a base level of detailed data.

Summary

- **Business Intelligence (BI)** refers to technology infrastructure for gaining maximum information from available data for the purpose of improving business processes. Typical BI infrastructure components are as follows: software solution for gathering, cleansing, integrating, analyzing and sharing data. Business Intelligence produces analysis and provides believable information to help making effective and high quality business decisions.
- The most common kinds of Business Intelligence systems are:
  - **EIS** - Executive Information Systems
  - **DSS** - Decision Support Systems
  - **MIS** - Management Information Systems
  - **GIS** - Geographic Information Systems
  - **OLAP** - Online Analytical Processing and multidimensional analysis
- **CRM - Customer Relationship Management :** Business Intelligence systems based on Data Warehouse technology. A Data Warehouse (DW) gathers information from a wide range of

company's operational systems, Business Intelligence systems based on it. Data loaded to DW is usually good integrated and cleaned that allows to produce credible information which reflected so called 'one version of the true'.

- **Business Intelligence tools:** The most popular BI tools on the market are:
  - Oracle - Siebel Business Analytics Applications
  - SAS - Business Intelligence
  - SAP - Business Objects XI
  - IBM - Cognos 8 BI
  - Oracle - Hyperion System 9 BI+
  - Microsoft - Analysis Services
  - MicroStrategy - Dynamic Enterprise Dashboards
  - Pentaho - Open BI Suite
  - Information Builders - WebFOCUS Business Intelligence
  - QlikTech - QlikView
  - TIBCO Spotfire - Enterprise Analytics
  - Sybase - InfoMaker
  - KXEN - IOLAP
  - SPSS - ShowCase
- **OLTP / OLAP**

  OLTP: On Line Transaction Processing which contains a normalized tables and online data which have frequent insert/updates/delete.
  Current data
  Short database transactions
  Online update/insert/delete
  Normalization is promoted
  High volume transactions
  Transaction recovery is necessary
  Since in OLTP, tables are normalized and hence query response will be slow for end user.
  OLAP: Online Analytical Programming contains the history of OLTP data is used for creating forecasting reports
  Current and historical data
  Long database transactions
  Batch update/insert/delete
  Demoralization is promoted
  Low volume transactions
  Transaction recovery is not necessary
- The roll up operation performs aggregation on data cube by climbing up a dimensional hierarchy.
- Drill down is reverse of roll up and can be realized by stepping down the dimensional hierarchy.
- The slice operation performs the selection on one dimension of given cube resulting in a sub cube.
- The dice operation defines a subcube by performing a selection on 2 or more dimensions.
- Pivot is a visualization operation that rotates the data axes in view in order to provide an alternative presentation of the data.

- **Approaches for OLAP SERVERS**

  Three possibilities for OLAP servers

  (1) **Relational OLAP (ROLAP):** Relational and specialized relational DBMS to store and manage warehouse data. OLAP middleware to support missing pieces.

  (2) **Multidimensional OLAP (MOLAP):** Array-based storage structures.
  Direct access to array data structures.

  (3) **Hybrid OLAP (HOLAP)**
  Storing detailed data in RDBMS
  Storing aggregated data in MDBMS
  User access via MOLAP tools.

Review Questions

1. Describe twelve rules given by CODD for multidimensional OLAP environment.
2. Discuss different models of OLAP.
3. Differentiate between MOLAP, ROLAP.
4. Explain different operations of an OLAP system.
5. Write a note on multidimensional data model.

❑❑❑

5

Chapter

Data Warehouse Implementation

INSIDE THIS CHAPTER

5.1. Introduction **5.2.** Materialization **5.3.** Cube Computation **5.4.** Complex Aggregation at Multiple Granularities: Multifeature Cube **5.5.** Testing and Tuning the Data Warehouse **5.6.** Query Manager **5.7.** Case Study: The Data Warehouse Manager **5.8.** Query Management Facility **5.9.** Future of Data Warehousing.

This chapter deals with data warehouse implementation issues. Since data warehouse contains large volumes of data and it is in demand that the queries must be answered in seconds. So in this chapter various techniques are studied for computing the data cube efficiently.

5.1. INTRODUCTION

As already said that OLAP servers demand that decision support queries must be answered in order of seconds, so data warehouse system must support highly efficient computation technology for data cubes, access methods and query processing techniques. How these methods and techniques can be applied to multidimensional data cubes will be explained in this chapter.

5.1.1. Efficient Data Cube Computation

In SQL the efficient computation of aggregations across many sets of dimensions are referred as "group by".

Data cube can be viewed as a lattice of cuboids. Therefore one of the approach for computation of cube is to use "compute cube" operator.

5.1.2. Lattice of Cuboids

Efficient data cube computation also depends upon how many cuboids are there in a cube. There are two cases, considered for this.

- The full or nothing case
- When dimension hierarchies are present (Fig. 5.1)

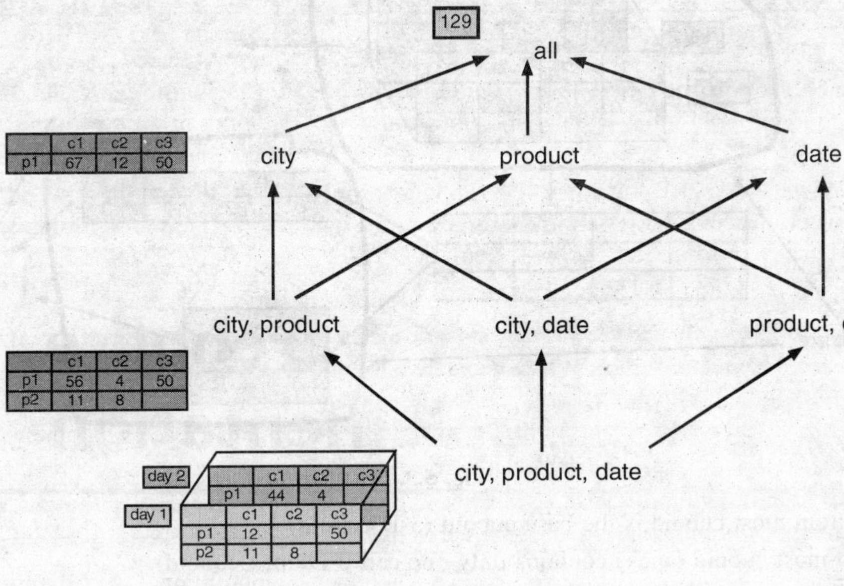

Fig. 5.1. Lattice of Cuboids.

Dimension Hierarchies

Secondly the efficient data cube computation also depends upon what is the size of each cuboid?

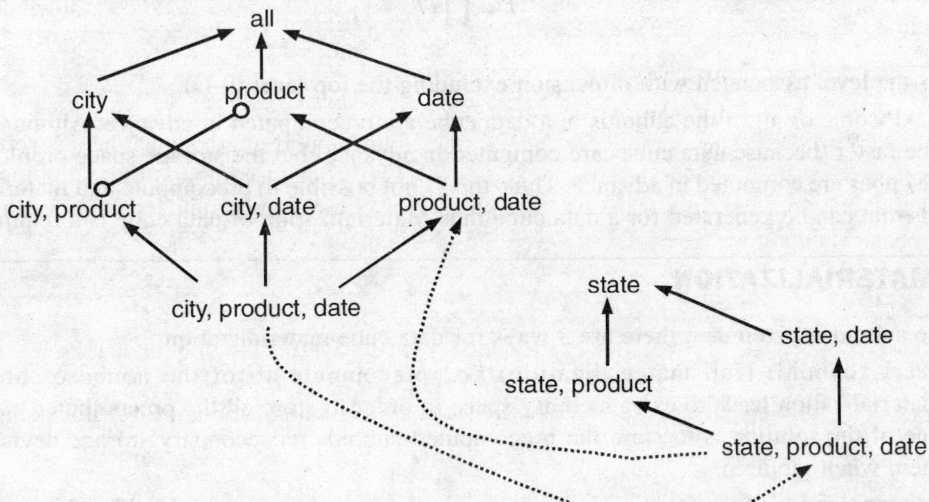

Fig. 5.2. Dimension Hierarchies.

Example:

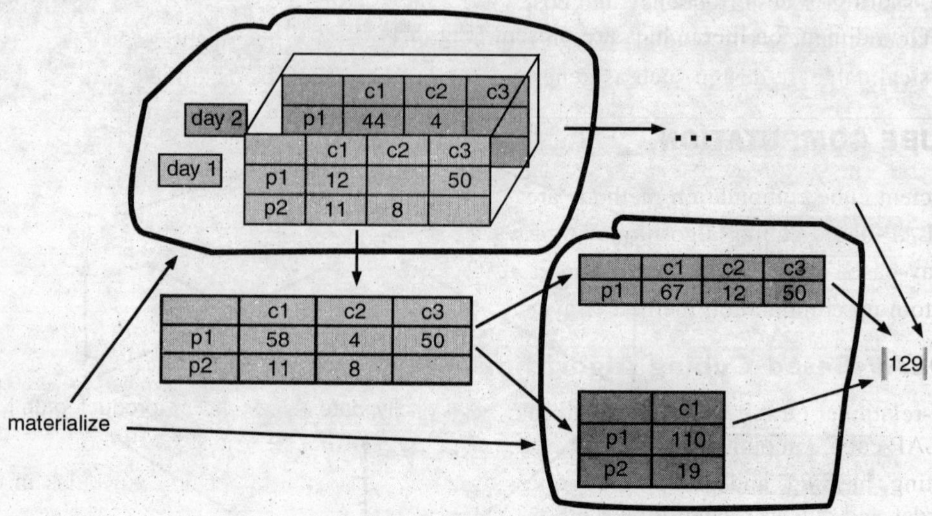

Fig. 5.3.

- The bottom-most cuboid is the base cuboid (3-D cuboid)
- The top-most cuboid (apex) contains only one cell(0-D apex cuboid)

But now the question is how many cuboids in an *n*-dimensional cube with L levels? The answer is for a cube of n dimension, there are a total of 2^n cuboids, including the base cuboid. But if there are no hierarchies exists with each dimension then the total number of cuboids for n-dimensional data cube is 2^n, but many dimensions have hierarchies like hours<day<weeks<month<quarter<year, therefore total number of cuboids for n-dimensional data cube is

$$T = \prod_{i=1}^{n} (L_i + 1)$$

where L is the level associated with dimension excluding the top level (0-D).

In OLAP some or all of the cuboids in a data cube are precomputed in advance. Although response time will be faster (because data cubes are computed in advance) but the storage space problem exists if all the data cubes are computed in advance. Thus this is not possible to precompute and materialize all of the cuboids that can be generated for a data cube thus Materialization of data cube is a requirement.

5.2. MATERIALIZATION

If base cuboid is given then there are 3 ways for data cube materialization

1. every (cuboid) (full materialization) *i.e.,* precompute all of the nonbase cuboids. Full materialization leads to extra memory space in order to store all the precomputed cuboids. But one of the solution is to store the precomputed cuboids on secondary storage devices and use them when required.
2. none (no materialization), *i.e.,* do not precompute all of the nonbase cuboids. No materialization leads to high cost as multidimensional aggregates are to be computed (because no one is precomputed).
3. some (partial materialization) *i.e.,* selectively compute a proper subset of the complete set of possible cuboids. As partial means neither full nor no materialization, therefore it presents a trade-off between storage space and response time.

The Selection of which cuboids to materialize is based on

1. size, sharing, access frequency and cost
2. workload characteristics, total storage requirements, cost for incremental updates.
3. Physical database design such as generation and indices (will be described in next topic).

5.3. CUBE COMPUTATION

The efficient cube computation methods are:

- ROLAP-based cubing algorithms (Agarwal et al'96)
- Array-based cubing algorithm (Zhao et al'97)
- Bottom-up computation method (Bayer & Ramarkrishnan'99)

5.3.1. ROLAP-Based Cubing Algorithms

ROLAP-relational OLAP uses tuples and relational tables. The following optimization technique is used in ROLAP cube computation.

- Sorting, hashing, and grouping operations are applied to the dimension attributes in order to reorder and cluster related tuples
- Grouping is performed on some subaggregates as a "partial grouping step"
- Aggregates may be computed from previously computed aggregates, rather than from the base fact table

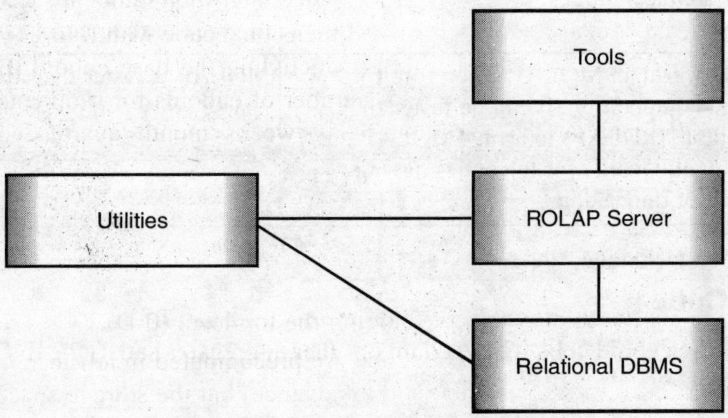

Fig. 5.4. ROLAP Based Cubing Algorithm.

5.3.2. MOLAP Base Cubing Algorithm

MOLAP –multidimensional OLAP is developed for array based cube construction of MOLAP. The following optimization technique is used in ROLAP cube computation.

1. **Partition the array into chunks:** Chunking is a method for dividing the n dimensional array into small n-dimensional chunks. The number of chunks depends upon the size of chunk. To remove the wastage space that results from empty array cells the chunks are compressed. Thus this technique is helpful in handling sparse cubes both on disk and in memory.
2. **Multiway array aggregation:** The order in which cells are visited can be optimized so as to minimize the number of times that each cell must be revisited. Thereby reducing the memory access and storage cost. Since the chunking technique involves the overlapping some of the aggregation computations, it is referred as "Multiway array aggregation".

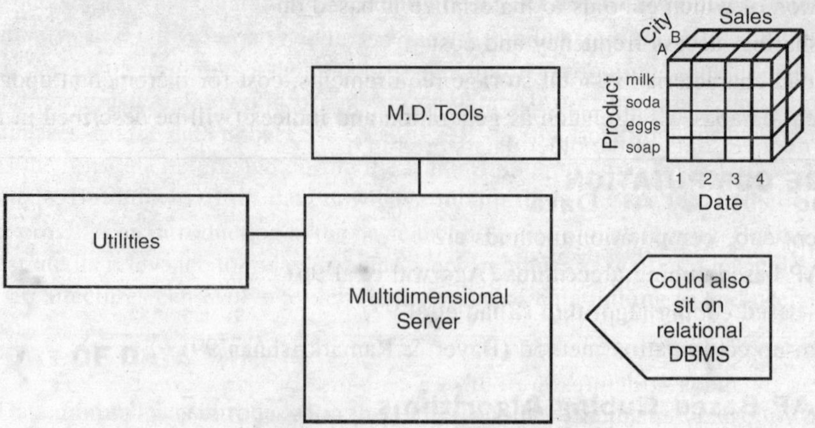

Fig. 5.5. Multi-dimensional OLAP Server.

Difference between ROLAP AND MOLAP based techniques

S.No.	ROLAP	MOLAP
1.	In ROLAP value based addressing is used where the dimension values are accessed by key-based addressing search techniques.	MOLAP Can't perform value based addressing, rather direct array addressing is used where dimension values are accessed via the position or index of their corresponding array position.
2.	ROLAP is slower than MOLAP. Therefore for ROLAP cube computation, it can be faster to convert the table (ROLAP) to array (MOLAP). But this technique works for the cubes having small number of dimensions.	MOLAP is faster than ROLAP because the sparse array compression technique is used.

Other computation techniques are:

5.3.3. Iceberg Cube

Computing only the cuboid cells whose count or other aggregates satisfying the condition like having count (*) >= minsup.

Fig. 5.6. Iceberg Cube.

Motivation behind ICEBERG CUBES is that

1. Only a small portion of cube cells may be "above the water" in a sparse cube
2. Only calculate "interesting" cells—data above certain threshold

Also incremental update methods have been developed to add in some new data to a precomputed cube or to remove some of the older data stored in it.

5.3.4. Indexing OLAP Data

Recently, data warehouse system is becoming more and more important for decision-makers. Most of the queries against a large data warehouse are complex and iterative. The ability to answer these queries efficiently is a critical issue in the data warehouse environment. With multidimensional data the indices help to reduce the overhead of scanning the extremely large tables.

If the right index structures are built on columns, the performance of queries, especially ad hoc queries will be greatly enhanced. Types of indexing techniques are:

1. Bitmap Indexes
2. Join Indexes

1. Bitmap index

It is a collection of bitmaps — one for each distinct value of the column. In this index technique, each bitmap has N bits where N is the number of rows in the table and a bit corresponding to a value v for a row r is set if and only if r has the value for the indexed attribute

Bitmap indices are:

Specially advantageous for low-cardinality domains

Represent each row of a table by a bit and the table as a bit vector

There is a distinct bit vector Bv for each value v for the domain

Example: the attribute sex has values M and F. A table of 100 million people needs 2 lists of 100 million bits

Gender	vote	Gender(M)	vote
M	Y	1	1
F	Y	0	1
F	N	0	0
M	N	1	0
F	Y	0	1
F	N	0	0

SQL query for this task can be written as

```
Query : select * from customer where
gender = 'M' and vote = 'Y'
```

The resultant is

```
1
0
0
0
0
0
```

Bitmap indecies facilitate easy functions such as join and aggregation through the use of bitmap operations. There is significant reduction in space as compared to more traditional indices where the pointers to the records have to be maintained. The products that support bitmaps are: Model204, oracle 7.3

2. Join Index

Join indices support precomputed joins. When used for multidimensional data a common approach is to create a join index between the dimensional table and a fact table. A join index between a fact table and a dimension table correlates a dimension tuple with the fact tuples that have the same value on the common dimensional attribute. Join indexes can be created using bitmaps as opposed to pointers.

For example, a join index on *city* dimension of *calls* fact table correlates for each city the calls (in the *calls* table) from that city.

Join indexes can also span multiple dimension tables

For example, a join index on *city* and *time* dimension of *calls* fact table

Star Join Processing

Use join indecies to join the fact and dimension table

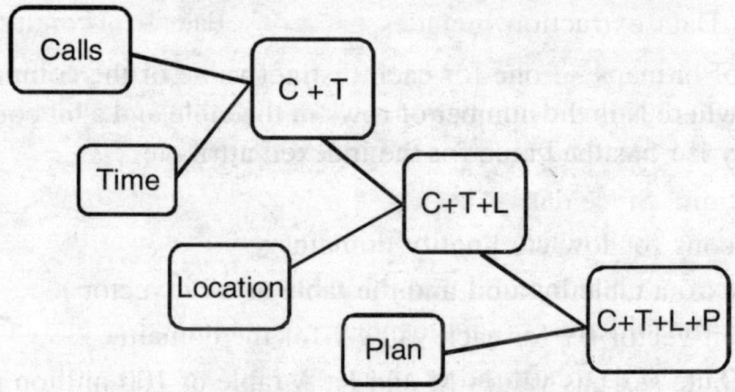

C = Call, T = Time, L = Location, P = Plan

Fig. 5.7. Star Join Processing.

The optimization can be achieved using select queries.

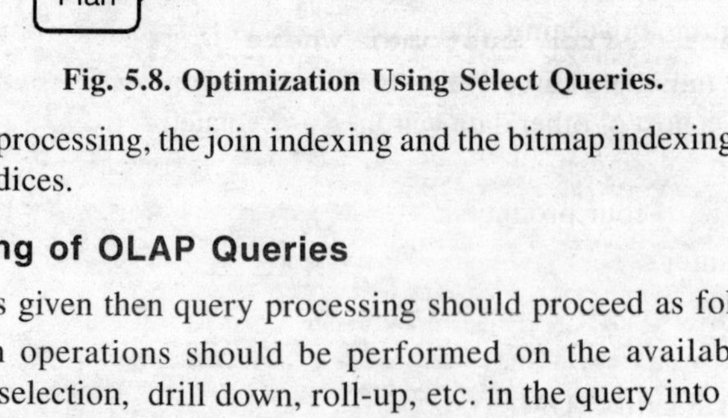

Fig. 5.8. Optimization Using Select Queries.

To speed up the query processing, the join indexing and the bitmap indexing methods can be integrated to form bitmapped join indices.

3. Efficient Processing of OLAP Queries

If materialized view is given then query processing should proceed as follows:

1. Determine which operations should be performed on the available cuboids: this includes transforming any selection, drill down, roll-up, etc. in the query into corresponding SQL and/or OLAP operations, e.g, dice = selection + projection

2. Determine to which materialized cuboid(s) the relevant operations should be applied: this involves the identification of all the materialized cuboid that is used to solve the qucrics, selecting the cuboid with the least cost etc.
3. Exploring indexing structures and compressed vs. dense array structures in MOLAP.
4. Data warehouse backend tools and utilities.

5.3.5. Data Warehousing Tools

Various Data warehousing Tools available in the market are the following :

- Informatica
- Cognos
- Business Objects
- Brio
- Oracle Warehouse builder.

Backend tools and utilities are used by Data warehouse systems to load and refresh the data. These tools and utilities include the following functions;

1. **Data extraction:** Data extraction includes gathering data from multiple, heterogeneous and external sources.
2. **Data cleaning:** Data cleaning deals with noisy data, missing data. This function is necessary to detect the errors in the data and rectifies it whenever needed.
3. **Data transformation:** Since data is collected from heterogeneous sources, data transformation function converts data from legacy or host format to warehouse format.
4. **Load:** After extraction, cleaning and transformation the data is loaded in the DW. Load which sorts, summarizes, consolidates, computer views check integrity and build indices etc.
   Load Techniques: Use SQL to append or insert new data
5. **Refresh:** Refreshing means are propagated updates from source data to the warehouse. Refreshing can be done either periodically (e.g., every night, every week) or after significant events or on every update: not warranted unless warehouse data require current data (up to the minute stock quotes) or refresh policy set by administrator based on user needs and traffic.
   Refresh Techniques: Fully Extract from base tables
   1. read entire source table: too expensive
   2. may be the only choice for legacy systems

for improving the quality of data cleaning and transformation are the important steps.

EXAMPLE: DB2 Products for data warehousing.

IBM's DB2 Warehouse Manager (DB2 WM) provides a distributed, heterogeneous infrastructure for designing, building, maintaining, governing, and accessing highly scalable, robust DB2 data warehouses.

DB2 WM helps in moving data directly from source to target and controls the server on which transformations — as well as a host of other data warehousing related actives — take place with distributed warehouse agents.

DB2 WM is really a suite of four products:

- Data Warehouse Center
- Information Catalog
- Query Patroller
- Query Management Facility (QMF)

5.4. COMPLEX AGGREGATION AT MULTIPLE GRANULARITIES: MULTIFEATURE CUBE (ROSS, et al. 1998)

The response of data mining queries is facilitated by data cubes because they allow the computation of aggregate data at multiple levels of granularity. The multifeature cubes compute complex queries involving multiple dependent aggregates at multiple granularities.

Features of Multifeature Cube

1. The multifeature cube graph helps in illustrating the aggregate dependencies in the query.
2. The complex data mining queries can be answered by multifeature cubes without any significant increase in the computational cost in comparison to the computation of cube for simple queries with standard data cubes.
3. The computation of a multifeature cube depends upon type of aggregate functions used in a cube. Two types of aggregate functions are:

   (*a*) **Distributive:** If the result derived by applying the function to *n* aggregate values is the same as that derived by applying the function on all the data without partitioning then called distributive. Examples of distributive aggregate functions are : count(), sum(), min(), max()

   (*b*) **Algebraic:** Aggregate functions said to be algebraic if it can be computed by an algebraic function with *M* arguments (where *M* is a bounded integer), each of which is obtained by applying a distributive aggregate function.

   For example, avg(), min_N(), standard_deviation().

   (*c*) **Holistic:** If there is no constant bound on the storage size needed to describe a subaggregate. Holistic function include, median(), mode(), rank().

Example 5.1. Grouping by all subsets of {item, region, month}, find the maximum price in 1998 for each group, and the total sales among all maximum price tuples

Solution. SQL syntax is

```
select item, region, month, max(price), sum(R.sales)
from purchases
where year = 1998
cube by item, region, month: R
such that S.price = max(price)
```

Description

1. In this the tuple representing purchase in 1998 is first selected.
2. the cube by clause computes the aggregation for all possible combinations of the attributes. Cube by is an n-dimensional generalization of group by clause. The attributes specified in the "cube by" clause are the grouping attributes. Tuples with the same value on all grouping attributes form one group.
3. S is the grouping variable ranging over the tuples in a group and whose price is equal to maximum in the group. This is specified in " such that" clause.
4. The sum of sales of the tuples in the group that S ranges over is computed and returned with the values of the grouping attributes.
5. The resultant cube is called" multifeature cube" in that it supports complex data mining queries for each multiple dependent aggregates are computed at a variety of granularities.

5.4.1. Other Methods

There are lots of methods exists for effective computation of data cubes. These are:

1. For distributive multifeature cube, the high level granularities are computed from lower-level granularities. As the above example is one of such type of case. The computation of max(price) at higher level of granularity group can be done by taking the maximum of all of the max(price) values at lower level granularities. Similarly sum (sales) can be computed.
2. Optimization techniques can be used based on estimated size of answers of groups within a data cube. There is no increase of I/O complexity because the output size for each group in multifeature group is constant and same estimation techniques can be used for intermediate results but the computation of hoslistic aggregate function is expensive than the distributive cubes thus CPU cost is increased.

Thus multifeature cubes can be used to answer the complex queries although there is addition of cost in comparison to simple data queries which is acceptable.

5.5. TESTING AND TUNING THE DATA WAREHOUSE [MDU Dec. 2011]

Testing the data warehouse and business intelligence system is critical to success. Without testing, the data warehouse could produce incorrect answers and quickly lose the faith of the business intelligence users. Effective testing requires putting together the right processes, people and technology and deploying them in productive ways. There are different levels of testing: unit testing, Integration testing, System testing, Acceptance testing, Performance testing. But still there are some challenges to DW testing:

5.5.1. Challenges of Data Warehouse Testing

- Data selection from multiple source systems and analysis that follows pose great challenge.
- Volume and the complexity of the data.
- Inconsistent and redundant data in a data warehouse.
- Loss of data during the ETL process.
- Non-availability of comprehensive test bed.
- Critical Data for Business.

Therefore the aim of testing the DW is:

Testing Goal: Our main aim is to check the quality of the data.

- Data completeness: Data completeness ensures that all expected data is loaded.
- Data transformation: Data transformation ensures that all data is transformed correctly according to business rules and/or design specifications.
- Data quality: Data quality ensures that the ETL application correctly rejects, substitutes default values, corrects or ignores and reports invalid data.

Data Warehousing Testing Responsibilities

Who should be involved with testing? The right team is essential to success:

- Business Analysts gather and document requirements
- QA Testers develop and execute test plans and test scripts
- Infrastructure people set up test environments
- Developers perform unit tests of their deliverables
- DBAs test for performance and stress
- Business Users perform functional tests including User Acceptance Tests (UAT).

Data Warehousing Test Plan

The Test Plan, typically prepared by the QA Testers, describes the tests that must be performed to validate the data warehousing and business intelligence system. It describes the types of tests and the coverage of required system features.

Test Cases are the details that enable implementation of the Test Plan. The Test Case itemizes steps that must be taken to test the system along with expect results. A Text Execution Log tracks each test along with the results (pass or fail) of each test item.

Different types of testing that are done while implementing the data warehouse are :

1. Unit testing: Developers perform tests on their deliverables during and after their development process. The unit test is performed on individual components and is based on the developer's knowledge of what should be developed.

Unit testing should definitely be performed before deliverables are turned over to QA by developers. Tested components are likely to have fewer bugs.

2. Integration testing: As the name indicates "integration", different modules that make up a component of DW application are tested together to ensure that they work together

3. System testing: In System testing the entire DW application is tested as a single unit to evaluate the system's compliance with its specified requirements. System testing falls within the scope of black box testing, and as such, should require no knowledge of the inner design of the code or logic.

4. Acceptance testing: This testing also tests the complete DW application but during this testing the users will conduct their own tests on the systems. Types of acceptance testing: Alpha and BETA testing, User acceptance testing, etc.

5. Performance testing: Performance testing can verify that a system meets the specifications claimed by its manufacturer or vendor. The process can compare two or more devices or programs in terms of parameters such as speed, data transfer rate, bandwidth, throughput, efficiency or reliability

These kinds of tools can facilitate testing and problem correction:

- Automated test tool
- Test data generator
- Test data masker
- Defect manager
- Automated test scripts.

Can the Data Warehouse Perform?

Tests can be designed and executed that show how well the system performs with heavy loads of data :

Extract Performance Test	Transform and Load Performance Test	Analytics Performance Test
Test the performance of the system when extracting a large amount of data.	Test the performance of the system when transforming and loading a large amount of data. Testing with a high volume is sometimes called a "stress test".	Test the performance of the system when manipulating the data through calculations.

5.5.2. Business Users Test Business Intelligence

Does the system produce the results desired by business users? The main concern is functionality, so business users perform functional tests to make sure that the system meets business requirements. The testing is performed through the user interface (UI) which includes data exploration and reporting.

Correctness Test: The system must be produce correct results. The measures and supporting context need to match numbers in other systems and be calculated correctly.

Usability Test: The system should be as easy to use as possible. It involves a controlled experiment about how business users can use the business intelligence system to reach stated goals.

Performance Test: The system must be able to return results quickly without bogging down other resources. Thus for a good DW testing is essential to ensure good quality of data.

5.5.3. Tuning The DATA WAREHOUSE

The general meaning of tuning is

(*i*) Just (a musical instrument) to the correct or uniform pitch or

(*ii*) Adjust (a receiver circuit such as a radio or television) to the frequency of the required signal.

Same is the case with DW *i.e.,* we know that DW usage will keep evolving over time as the user profile changes. The change in user profiles and data access patterns has a direct impact on the queries that are being run. These changes will affect the aggregation tables that need to be created. Some of the measures to know why tuning is required are:

1. Average query response times and elapsed times for various queries, and
2. I/O throughput rates.
3. Memory requirement for every process.

The ad hoc nature of data warehousing restricts the amount of tuning that can be done. However there are some tuning strategies that can be considered.

5.5.4. Tuning Strategies

Following are major areas to focus on when tuning the data warehouse:

- hardware considerations,
- data load tuning,
- query tuning.

1. Hardware Consideration

As the number of users and the size of the data warehouse increases there is a need of additional hardware that can help in improving performance,.

For example:

- increase CPU bandwidth,
- increase number of disks to spread I/O processing,
- add controllers if the disk controller bandwidth is exceeded.

An understanding of where problems exist is needed before assuming that any additional hardware will improve performance. For example, obtaining additional CPU bandwidth will not solve performance problems if the problems are caused by I/O bottlenecks (in this situation, increasing the CPU bandwidth may even worse the bottleneck).

2. Data Load Tuning in the Data Warehouse

We know that the data warehouse cannot be used until data load is completed. Data loading is a critical component and is done during overnight processing and serious effects can be seen if the data load is not completed in its assigned window. Therefore following things can be done to improve the data load performance.

(*a*) **Maintaining Indexes:** One potentially costly area of the data load is the maintenance of indexes on data being loaded. This is particularly costly if data is being loaded to tables that already contain data.

Indexes can either be maintained while data is being loaded or they can be dropped and then rebuilt once the data load is completed. Dropping indexes may be quicker, especially if the database management system allows indexes to be rebuilt in parallel.

Even if it takes longer to load the data and then rebuild indexes, it may still be the preferred strategy if it allows additional processing to begin earlier (e.g., begin the next step in the overnight processing once the data has been loaded; even if the indexes have not been rebuilt).

***(b)* Integrity Checking:** Another area that can have significant impact on the data load process is the amount of integrity checking.

Try to keep the number of integrity checks to be applied to the data warehouse to a minimum. Wherever possible, perform integrity checks on the source systems.

Examine the integrity checking that is being conducted to determine if any improvements can be identified.

Other Considerations: Loading large quantities of data is I/O intensive and can be CPU intensive, especially if numerous checks and transformations are being applied to each record being loaded. To improve load speeds direct loads can be used (e.g., place data into pre-formatted blocks, which are then written to the database), Multiple parallel processors can be used (*i.e.,*, use multiple processes to spread the CPU load amongst available CPUs), however while doing these care must be taken to avoid I/O bottleneck. Then The source files can then be spread over multiple disks to avoid I/O bottlenecks and contention on a single disk.

Finally care must be taken to see that the load destination does not become a bottleneck especially when a large number of data processes are being used, This can be achieved by spreading the database files across multiple disks or by spreading the table to be loaded over multiple database files.

3. Query Tuning

There are two types of queries in the data warehousing environment.

(*a*) Fixed queries are clearly defined and well understood.

(*b*) ad hoc queries are unpredictable; in both frequency and quantity. Different techniques are applied to tune each type of query.

***(a)* Fixed Queries:** Tuning the Fixed queries is done in the same manner as the traditional databases are tuned. Fixed queries include pre-defined queries and reports that are routinely executed and These queries have predictable behaviour. The most successful execution plan is retained when testing fixed queries.

The only real parameter with the fixed queries is the size of data being queried. Even this is also not an issue as such because the query can be tested as often as desired.

Little can be done to fact data to improve performance, however, dimension data may be furthered tuned by creating additional aggregations and/or indexes.

***(b)* Ad Hoc Queries:** The number of users of the DW have a important effect on the performance of the data warehouse, especially the number of users who perform ad hoc queries. The team must have a clear understanding of the usage profiles of all the users. If there is clear understanding of the usage profiles of users then it will help to make the query from ad hoc queries more predictable. Since the usage profiles keep on changing it is necessary to obtain the following information on users/user groups:

1. the number of different user groups that were previously identified when user access requirements were defined,
2. number of users in each group,
3. frequency and patterns of ad hoc querying,
4. average and maximum query sizes,

5. peak usage times in daily cycle,
6. peak usage days in weekly/monthly cycles,
7. query volume during peak times.

Understanding usage profiles does not provide any specific tuning solutions, however, it may improve the ability to anticipate the types and number of ad hoc queries to be executed.

(c) Processing Queues

One way to control the query load is to implement batch queues. Query services can control the queries that are submitted by placing the queries in queues. Queries are then executed when they work their way to the front of the queue. The intent is to manage the processing environment so that a flood of queries does not cripple the system's ability to provide reasonable performance.

Once a queuing strategy is implemented, determine how many queues are to be used; *i.e.*,, how many queries are allowed to run concurrently?

(d) Other Considerations

Queries are most efficient when run against aggregations instead of directly against the fact data. Analyze ad hoc queries and create additional aggregations when appropriate.

Drop indexes and aggregations that are no longer used.

Decisions can best be made by analyzing ad hoc queries. Use query services to collect and analyze the following information on each query: i.e query structure, query execution plan, elapsed execution time of query, frequency with which query is executed,· I/O, CPU, memory usage.

5.6. QUERY MANAGER

The query manager performs all operations associated with management of user queries. This component is usually constructed using vendor end-user access tools, data warehousing monitoring tools, database facilities and custom built programs. The complexity of a query manager is determined by facilities provided by the end-user access tools and database.

1. One of the job of query manager is to prioritize the query if 2 or more queries run simultaneously. If two or more queries run simultaneously then problem can occur in the system, because these query tend to monopolize that is wish to use all the resources. Thus this monopolization of the system should be stopped and if such query once detected then it should be killed at that time. The solution of this problem is that the queries must be modified so that it runs in a particular amount of time using reasonable resources.
2. The other job of the query manager is to collect the data.
3. Maintenance of following data is also job of query manager:
   (*a*) Syntax of the query
   (*b*) Processor used
   (*c*) I/O devices used.
   (*d*) Execution of the query
   (*e*) Frequency of execution of the query.
   (*f*) Usage of memory
   (*g*) Query elapsed time.
4. The transformation of query is the function of query manager. Query manager should be aware of aggregate tables.
5. If there are small and large queries to be run then the preference is given to small queries. But this may lead to starvation. This can be resolved by prioritization.

5.7. CASE STUDY: THE DATA WAREHOUSE MANAGER

The Data Warehouse Manager supports a wide variety of relational and non-relational data sources. DB2 Universal Database warehouse can be filled with data from the most common relational databases (like any member of the DB2 family, Oracle, Sybase, Informix, Microsoft SQL Server), flat files, non-relational data stores, and 40 other data sources!

DB2 WM includes over 100 built-in transformers that you can use to cleanse, restructure, correlate, standardize, and summarize data to make it more accurate and usable.

DB2 WM adds the following features to DB2 for data warehousing:

- Increase the scalability of data warehouses
- Fast deployment of data marts
- Comprehensive manageability and resource controls for DBAs
- Easy access to data and metadata by end users
- Enterprise reporting to develop and deliver reports to unlimited clients

Currently the DB2 WM is generally available for: AIX, Windows NT, Windows 2000, Solaris, IBM OS/2, zSeries, and iSeries.

5.7.1. The Data Warehouse Center

DB2 on Windows comes with the IBM Data Warehouse Center. This tool offers a subset of the features of the Data Warehouse Manager for Windows-based databases. It is free of charge and allows DBAs a chance to sample some of the powerful features that they can leverage with Data Warehouse Manager.

The Data Warehouse Center offers:

- Registration and access to disparate data sources
- Definition of data extraction and transformation steps
- Population and the automation of warehouse management processes

The Data Warehouse Center supports full refresh and incremental update data movement options including leveraging the power of IBM's integrated data replication functions. The Warehouse Launchpad simplifies the job of populating a warehouse by leading you through the related tasks.

The Data Warehouse Center is a selectable component of a DB2 installation. If you select this component, the Control database will be created to help manage your data warehouse.

Query Patroller

DB2 Query Patroller provides DBAs with a facility for query governing and workload management. Essentially, Query Patroller acts as a police force that can limit the resource consumption of a query, based on a user's profile. Management tools like this are becoming a necessity in data warehousing with the proliferation of ad-hoc query tools that allows you to ask business questions of that data (which isn't always the most efficient SQL).

Information Catalog

Information Catalog empowers DBAs with a rich and powerful toolset that allows them to build, define, and administer metadata information. Metadata is really information about the information. Think of metadata as descriptors of the data. For example, if an aggregation was created in the data warehouse that summarized sales for root beer, a metadata entry might relate this information to the business analyst.

The effective management of metadata adds value in several ways, including by:

- Making the integration of solutions easier and freeing skilled resources for other Projects

- Improving impact analysis to rapidly respond to changing operational systems and changing business requirements
- Helping you to find and understand the data they need for decision making

5.8. QUERY MANAGEMENT FACILITY

Query Management Facility (QMF) for Windows is included with DB2 WM to provide a multipurpose query tool for business reporting, data sharing, server resource protection, robust application development, and native connectivity to all of the DB2 workstation platforms.

It allows the user to:

- Build queries and reports easily via its quick start interface.
- Use the new Java-based query capability to launch queries from your favourite browser.
- Easily integrate query results with desktop tools such as spread sheets and personal databases.
- Rapidly build data access and update applications.
- Fully exploit DB2 performance, SQL syntax and advanced database performance techniques such as static SQL.

5.9. FUTURE OF DATA WAREHOUSING

Data Warehousing is a new emerging field that it is difficult to estimate what new developments are likely to most affect it. Parallel DB servers with improved query engines are likely to be introduced. Parallel servers will make it possible to access huge data bases in much less time. Another new technology is data warehouses that allow for the mixing of traditional numbers, text and multi-media. The availability of improved tools for data visualization (business intelligence) will allow users to see things that could never be seen before. Over the next few years, the growth of data warehousing is going to be enormous with new products and technologies coming out frequently. In order to get the most out of this period, it is going to be important that data warehouse planners and developers have a clear idea of what they are looking for and then choose strategies and methods that will provide them with performance today and flexibility for tomorrow.

Summary

- Data generalization is a process that abstracts a large set of task-relevant data in a database from a relatively low conceptual level to higher conceptual levels.
- Challenges in MOLAP are:
  (*a*) Storing large arrays for efficient access
  (*b*) Row-major, column major
  (*c*) Chunking
  (*d*) Compressing sparse arrays
  (*e*) Creating array data from data in tables
  (*f*) Efficient techniques for Cube computation
- Bitmap Index:
  (*a*) Index on a particular column
  (*b*) Each value in the column has a bit vector: bit-op is fast
  (*c*) The length of the bit vector: # of records in the base table
  (*d*) The *i*-th bit is set if the *i*-th row of the base table has the value for the indexed column
  (*e*) not suitable for high cardinality domains

- Materialization Factors:
  - (*a*) Type/frequency of queries
  - (*b*) Query response time
  - (*c*) Storage cost
  - (*d*) Update cost
- Data Warehouse Back-End Tools and Utilities are data extraction, data cleaning, data transformation, data loading and refreshing.
- Hash/sort based methods (Agarwal et. al. VLDB'96)
  - (*a*) **Smallest-parent:** Computing a cuboid from the smallest cuboid previously computed cuboid.
  - (*b*) **Cache-results:** Caching results of a cuboid from which other cuboids are computed to reduce disk I/Os.
  - (*c*) **Amortize-scans:** Computing as many as possible cuboids at the same time to amortize disk reads.
  - (*d*) **Share-sorts:** Sharing sorting costs cross multiple cuboids when sort-based method is used.
  - (*e*) **Share-partitions:** Sharing the partitioning cost cross multiple cuboids when hash-based algorithms are used.
- Efficient Processing OLAP Queries:
  - (*a*) Determine which operations should be performed on the available cuboids:
  - (*b*) transform drill, roll, etc. into corresponding SQL and/or OLAP operations, *e.g.*, dice = selection + projection
  - (*c*) Determine to which materialized cuboid(s) the relevant operations should be applied.
  - (*d*) Exploring indexing structures and compressed vs. dense array structures in MOLAP.

8. Discovery-Driven Exploration of Data Cubes:

(*a*) Hypothesis-driven: exploration by user, huge search space

(*b*) Discovery-driven (Sarawagi et. al.'98)

1. Pre-compute measures indicating exceptions, guide user in the data analysis, at all levels of aggregation.
2. Exception: significantly different from the value anticipated, based on a statistical model.
3. Visual cues such as background color are used to reflect the degree of exception of each cell.
4. Computation of exception indicator (modeling fitting and computing SelfExp, InExp, and PathExp values) can be overlapped with cube construction.

Review Questions

1. How many cuboids are there in n-dimensional data cube? Explain.
2. Explain the need of materialization.
3. What are the types of materialization explain with their features ?
4. What are the issues in MOLAP ?
5. Which one is faster ROLAP or MOLAP and why?
6. How indexing is done ?
7. Explain bit index and join index.
8. Which method is used if new data to be added or removed from precomputed cube.
9. Explain efficient processing of OLAP queries.
10. Explain multifeature cubes.

□□□

6 Chapter

Introduction to Data Mining

INSIDE THIS CHAPTER

6.1. Overview **6.2.** History of Data Mining **6.3.** Defining Data Mining **6.4.** Basic Data Mining Tasks **6.5.** KDD Vs. Data Mining **6.6.** Data Mining Application **6.7.** The Scope of Data Mining **6.8.** Data Mining Techniques **6.9.** Why Use Data Mining? **6.10.** Data Mining Tools Comparison **6.11.** Advantages and Disadvantages of Data Mining **6.12.** Applications of Data Mining **6.13.** KDD-Data Mining Issues and Challenges

6.1. OVERVIEW

Data mining is a powerful new technology with great potential to help companies focus on the most important information in the data they have collected about the behaviour of their customers and potential customers. It discovers information within the data that queries and reports can't effectively reveal.

Data mining, also known as Knowledge Discovery in Databases (KDD), is the process of automatically searching large volumes of data for patterns. In order to achieve this, data mining uses computational techniques from statistics, machine learning, and pattern recognition.

Data mining tools predict future trends and behaviours, allowing businesses to make proactive, knowledge-driven decisions. Most companies already collect and refine massive quantities of data. Data mining techniques can be implemented rapidly on existing software and hardware platforms to enhance the value of existing information resources, and can be integrated with new products and systems as they are brought on-line.

Data mining is often confused with "writing reports and queries," when in fact data mining activities do not involve any traditional report writing or querying at all. Data mining is performed through a specialized tool, which executes predefined data mining operations based on analytical models. Data mining is the analysis of data with the intent to discover gems of hidden information in the vast quantity of data that has been captured in the normal course of running the business.

There is a difference between DBMS and Data mining. Traditional databases queries access a database using a well defined query stated in a language such as SQL. The output of the query is a subset of the databases but it may contain aggregation. Data mining access of a database differs from this traditional access in several ways like DBMS is a system for housing and managing a set of digital databases.

On the other hand data mining is a technique or concept in computing, which deals with the fact to extract useful and previously unknown information raw data. Most of the times, these raw data are kept in very large databases. So the data miners use existing DBMS functionalities for handling, directing and even pre-process the raw data before and during the Data Mining process. Yet, a DBMS alone can not be used to analyze data. But a few DBMS data now have inbuilt tools or capabilities that can analyze data

This chapter provides an introduction to the basic technologies of data mining. Examples of profitable applications illustrate its relevance to today's business environment as well as a basic description of how data warehouse architectures can evolve to deliver the value of data mining to end users.

6.2. HISTORY OF DATA MINING

The term "Data mining" was introduced in the 1990s, but data mining is the evolution of a field with a long history. This evolution began when business data was first stored on computers, continued with improvements in data access, and more recently, generated technologies that allow users to navigate through their data in real time.

Data mining roots are traced back along three family lines: classical statistics, artificial intelligence, and machine learning.

Statistics are the foundation of most technologies on which data mining is built, *e.g.,* regression analysis, standard distribution, standard deviation, standard variance, discriminant analysis, cluster analysis, and confidence intervals. All of these are used to study data and data relationships.

Artificial intelligence, or AI, which is built upon heuristics as opposed to statistics, attempts to apply human-thought-like processing to statistical problems. Certain AI concepts which were adopted by some high-end commercial products, such as query optimization modules for Relational Database Management Systems (RDBMS).

Machine learning is the union of statistics and AI. It could be considered an evolution of AI, because it blends AI heuristics with advanced statistical analysis. Machine learning attempts to let computer programs learn about the data they study, such that programs make different decisions based on the qualities of the studied data, using statistics for fundamental concepts, and adding more advanced AI heuristics and algorithms to achieve its goals.

Since Data mining is a natural development of the increased use of computerized databases to store data and provide answers to business analysts. From the user's point of view, the four steps listed in Table 6.1 were revolutionary because they allowed new business questions to be answered accurately and quickly.

Table 6.1. Evolution of data Mining

Evolutionary Step	Business Question	Enabling Technology
Data Collection (1960s)	"What was my total revenue in the last six years?"	Computers, tapes, disks
Data Access (1980s)	"What were unit sales in England last March?"	Faster and cheaper computers with more storage, relational databases
Data Warehousing and Decision Support	"What were unit sales in England last March? Drill down to Boston."	Faster and cheaper computers with more storage, On-line analytical processing (OLAP), multidimensional databases, data warehouses
Data Mining	"What's likely to happen to Boston unit sales next month? Why?"	Faster and cheaper computers with more storage, advanced computer algorithms

Data mining, in many ways, is fundamentally the adaptation of machine learning techniques to business applications. Data mining is best described as the union of historical and recent developments in statistics, AI, and machine learning. These techniques are then used together to study data and find previously-hidden trends or patterns within.

6.3. DEFINING DATA MINING

There are many definitions of the term "data mining" but the core concept behind all of them is the same. Data mining refers to extracting or mining the knowledge from large amount of data but this definition seems to be not completed.

One of the definition might be :" *The process of extracting previously unknown, valid and actionable information from large databases and then using the information to make crucial business decisions.*"

OR

"The science of extracting useful information from large data sets or databases."

- "Data mining is the search for relationships and global patterns that exist in large databases but are `hidden' among the vast amount of data, such as a relationship between patient data and their medical diagnosis. These relationships represent valuable knowledge about the database and the objects in the database and, if the database is a faithful mirror, of the real world registered by the database." *Marcel Holshemier and Arno Siebes (1994).*
- *The analogy with the mining process is described as* "Data mining refers to 'using a variety of techniques to identify nuggets of information or decision-making knowledge in bodies of data, and extracting these in such a way that they can be put to use in the areas such as decision support, prediction, forecasting and estimation. The data is often voluminous, but as it stands of low value as no direct use can be made of it is the hidden information in the data that is useful'." *Clementine User Guide, a data mining toolkit from SPSS.*
- "Data Mining is the nontrivial extraction of implicit, previously unknown, and potentially useful information from data. This encompasses a number of different technical approaches, such as clustering, data summarization, learning classification rules, finding dependency net works, analyzing changes, and detecting anomalies." *William J Frawley, Gregory Piatetsky-Shapiro* and *Christopher J Matheus.*

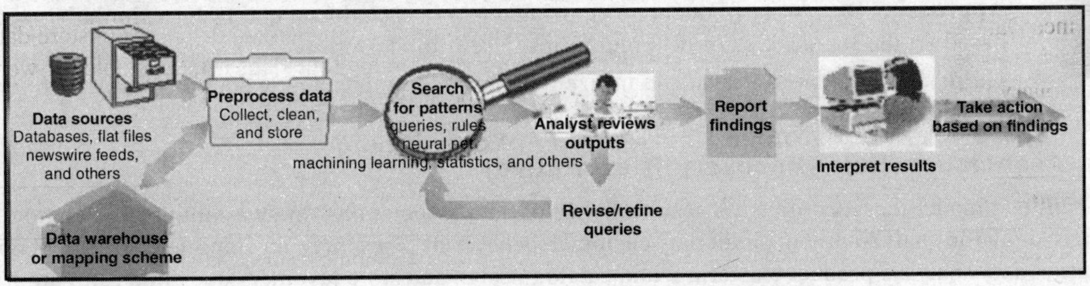

Fig. 6.1. Data Mining Process

Meaning of Unknown, Valid and actionable information: The information is previously unknown in that it is not directly derived from the data. Instead, the information takes the form of relationships among the database columns where the value in one or more columns predicts the outcome in another, hence the name, predictive model. But predictive models must be valid. Rating a model's predictive power usually involves testing it against another data set. The information is actionable, which suggests that there should be some goal in mind before putting any effort into data mining.

6.4. BASIC DATA MINING TASKS [MDU May 2012]

The two "high-level" primary goals of data mining, in practice, are *prediction* and *description*.

6.4.1. Prediction

Prediction involves using some variables or fields in the database to predict unknown or future values of other variables of interest. Example: to Judge if a patient has specific disease based on his/her medical tests results.

6.4.2. Description

Descriptive tasks focuses on finding human-interpretable patterns describing the data. Example: Identifying web pages that are accessed together. (human interpretable pattern).

The goals of prediction and description are achieved by using the following primary **data mining tasks**:

1. Classification [Predictive]

Classification maps (classifies) a data item into one of several predefined classes. Pattern recognition is a type of classification where an input pattern is classified into one of the several classes based on the similarity to these predefined classes.

2. Clustering [Descriptive]

Clustering is a common descriptive task where one seeks to identify a finite set of categories or clusters to describe the data.

Closely related to clustering is the task of *probability density estimation* which consists of techniques for estimating, from data, the joint multi-variate probability density function of all of the variables/fields in the database.

Example: Document Clustering:

- **Goal:** To find groups of documents that are similar to each other based on the important terms appearing in them.
- **Approach:** To identify frequently occurring terms in each document. Form a similarity measure based on the frequencies of different terms. Use it to cluster.
- **Gain:** Information Retrieval can utilize the clusters to relate a new document or search term to clustered documents.

3. Association Rule Discovery [Descriptive]

It is a model that identifies the specific types of data associations. In data mining, association rules are useful for analyzing and predicting customer behaviour. They play an important part in shopping basket data analysis, product clustering, and catalog design and store layout. Association rules are if/then statements that help uncover relationships between seemingly unrelated data in a relational database or other information repository. An example of an association rule would be "If a customer buys a dozen eggs, he is 80% likely to also purchase milk."

4. Regression [Predictive]

Regression is learning a function which maps a data item to a real-valued prediction variable. A regression task begins with a data set in which the target values are known. For example, a regression model that predicts children's height could be developed based on observed data for many children over a period of time. The data might track age, height, weight, developmental milestones, family history, and so on. Height would be the target, the other attributes would be the predictors, and the data for each child would constitute a case.

In the model build (training) process, a regression algorithm estimates the value of the target as a function of the predictors for each case in the build data. These relationships between predictors and target are summarized in a model, which can then be applied to a different data set in which the target values are unknown.

Regression models are tested by computing various statistics that measure the difference between the predicted values and the expected values.

5. Sequential Pattern Discovery [Descriptive]

Sequence discovery is used to determine the sequential patterns in data. These patterns are based on time sequence of actions for example most of the time it is analysed by the web masters how pages are accesses whether in the sequence or random etc. Depending on that links are provided from one page to another as we can see in goggle page are shown in the sequence and we can jump on any page.

6. Summarization

Summarization involves methods for finding a compact description for a subset of data. It is also called characterization or generalization. For example one of the criteria used to compare the universities is Ranking. This is a summarization that is helpful for the students who want to take admissions.

7. Dependency Modeling

Dependency Modeling consists of finding a model which describes significant dependencies between variables. Dependency models exist at two levels: structural level and quantitative level.

1. The *structural* level of the model specifies (often graphically) which variables are locally dependent on each other, and
2. The *quantitative* level of the model specifies the strengths of the dependencies using some numerical scale.

8. Change and Deviation Detection :

Change and Deviation Detection focuses on discovering the most significant changes in the data from previously measured values.

6.5. KDD VS. DATA MINING [MDU Dec. 2011]

Although KDD and Data Mining are used interchangeably but it is seen that KDD consists of many steps and Data mining is only one of the step.

Difference between KDD Vs. Data Mining

S. No.	KDD	Data Mining
1.	It is the process of finding useful information and patterns in data.	It is the use of algorithms to extract the information and patterns derived by the KDD process.
2.	KDD is the overall process of extracting knowledge from data	Data Mining is a step inside the KDD process, which deals with identifying patterns in data.

The steps involved in KDD are:

1. Data Selection

In this step the data is obtained from various databases, files etc. The data sources are different and may be heterogeneous.

2. Data Preprocessing

Data preparation is the key to KDD process. A prerequisite for any data mining algorithm is a set of selective, clean and transformed data. It is estimated that more than 50% of work in KDD is in getting the data to the point where data mining tools can actually start running. The business must have the understanding of what preparations are necessary for a data mining analysis.

The following figure displays the typical activities in Data Preparation.

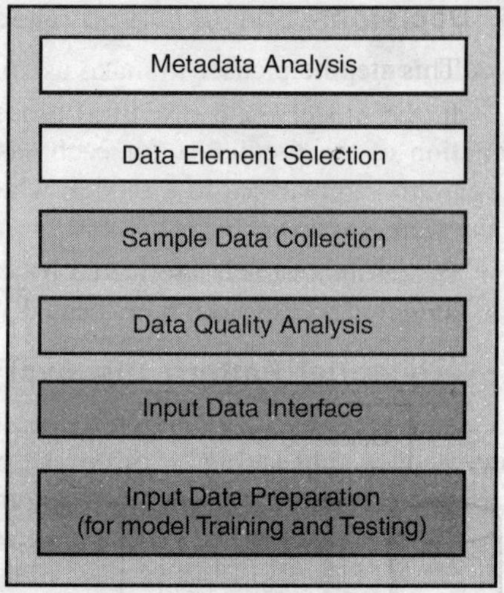

Fig. 6.2. Activities in Data Preparation.

Metadata is the most important component of a Business Intelligence system. KDD also highly relies on the availability and quality (consistency, accuracy and sufficiency) of metadata. For ensuring use of appropriate data elements from source, system metadata needs to be analyzed from quality perspective. Also business rules for integration and transformation need to be captured in form of process metadata to facilitate the Data Preparation. With the help of the metadata "all of the possibly required" input data elements are selected from various source systems.

After selecting the data elements from various sources, it's necessary to analyze the quality of the data. For this, data samples are collected from the sources and data profiling is performed to understand the Physical Data Quality issues. The Logical and Unmanaged/Unstructured Data Quality issues will need to be captured from the business users. The outcome of Data Quality Analysis will primarily aid in building the integration, transformation and cleansing rules repository (*i.e.,* Process Metadata).

Input Data Interface will need to be developed using the selected data elements and the process metadata. These interfaces will be used to fetch data from the source systems in required format. The data acquired from the sources will be needed for three prime purposes during Data Mining process *i.e.,* Training the data mining model, Testing the finalized model and Applying the model on complete data.

3. Data Transformation

The data even after cleaning are not ready for mining as we need to transform them into forms appropriate for mining. The techniques used to accomplish this are smoothing, aggregation, normalization etc.

4. Data Mining

Now we are ready to apply data mining techniques on the data to discover the interesting patterns. Techniques like clustering and association analysis are among the many different techniques used for data mining.

5. Pattern Evaluation and Knowledge Presentation

This step involves visualization, transformation, removing redundant patterns etc. from the patterns we generated. This step basically deals with how the data mining results are presented to the users. Visualization techniques include graphical, geometric, Icon based techniques etc.

6. Decisions/Use of Discovered Knowledge

This step helps user to make use of the knowledge acquired to take better decisions.

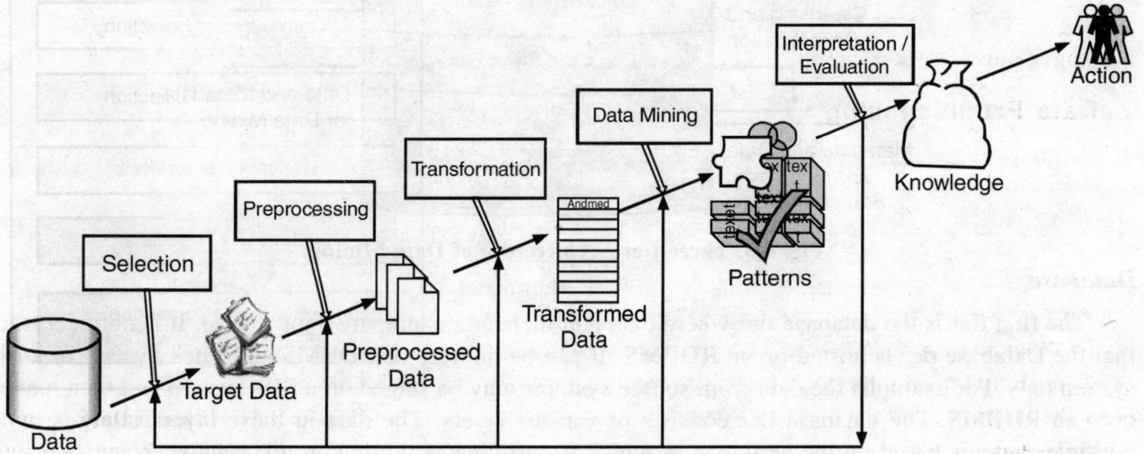

Fig. 6.3. Steps of Data Mining.

Based on this view, the architecture of data mining systems may have the following components shown in Fig. 6.4.

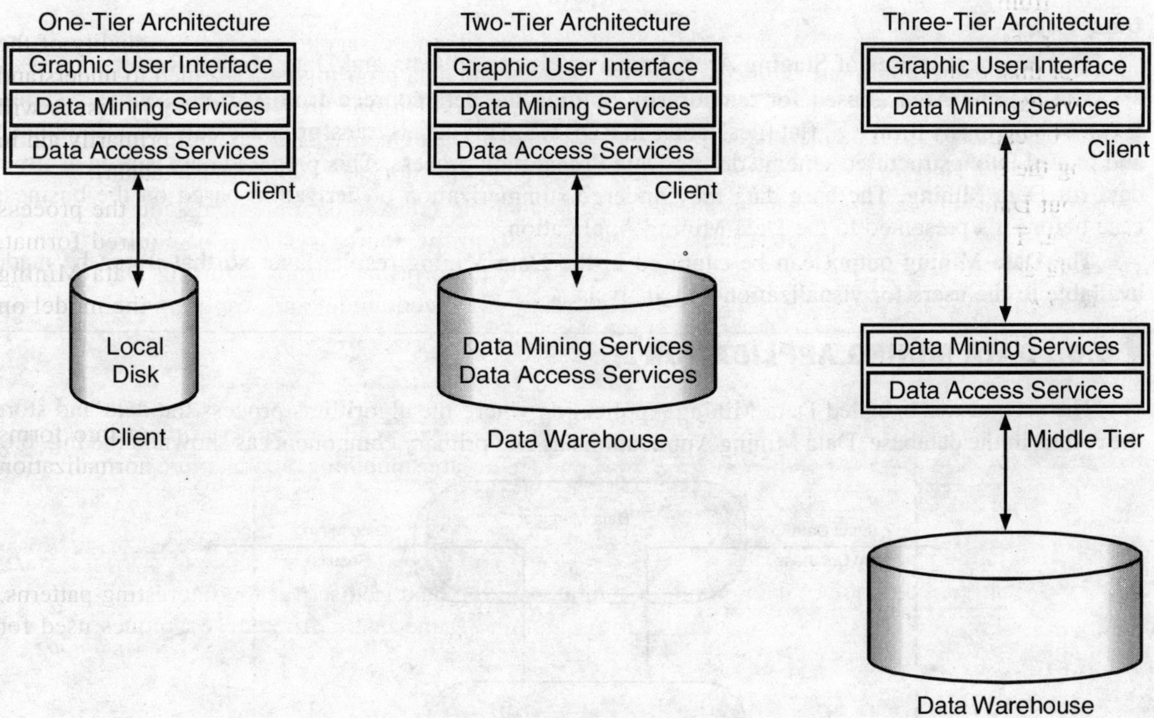

Fig. 6.4. Architecture of Data Mining Systems.

Generic three-tier architecture for Data Mining is explained in Fig. 6.5.

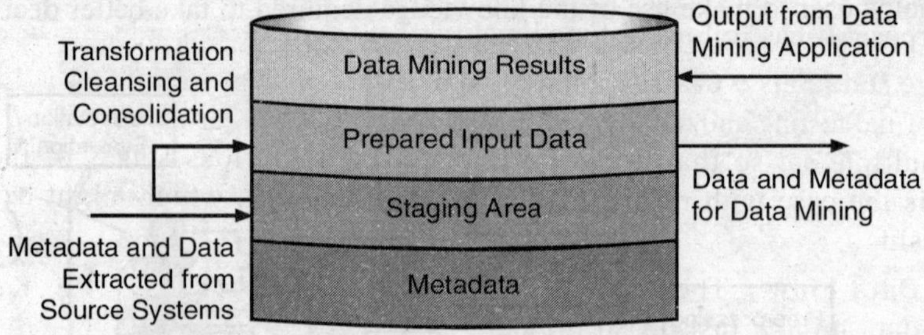

Fig. 6.5. Three-tier Architecture of Data Mining.

Database

The first tier is the database tier where data and metadata is prepared and stored. It is not necessary that the Database tier is hosted on an RDBMS. It can be mixture of RDBMS and Files System or a file system only. For example, the data from source systems may be staged on a files system and then loaded onto an RDBMS. The database tier consists of various layers. The data in these layers interface with multiple systems based on the activities in which it participates. Following diagram represents various layers in the Database tier.

Metadata Layer

The Metadata layer is the common and most frequently used layer. It contains information about sources, transformations and cleansing rules and the data mining results. It forms the backbone for the data in entire Data Mining Architecture.

Data Layer

This layer comprises of Staging Area, Prepared/Processed Data and Data Mining Results.

The Staging Area is used for temporarily holding the data sourced from various source systems. It can be held in any form *e.g.,* flat files, tables in RDBMS. This data is transformed, cleansed, consolidated and loaded into a structured schema during Data Preparation process. This prepared data is used as input data for Data Mining. The base data may undergo summarization or derivation based on the business case before it's presented to the Data Mining Application.

The Data Mining output can be captured in the Data Mining results layer so that it can be made available to the users for visualization and analysis.

6.6. DATA MINING APPLICATION

The second tier is called Data Mining Application where the algorithms process the data and store the results in the database. Data Mining Application has two primary components as shown in the Fig. 6.6.

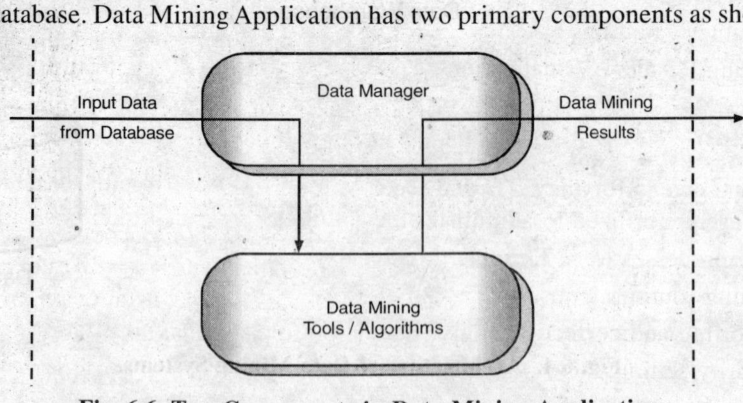

Fig. 6.6. Two Components in Data Mining Applications.

1. Data Manager

As the name suggests, this layer manages the data in the Database Tier and controls the data flow for data mining purpose. It has following functionality.

- **Manage Data Sets :** Classification of input data will be necessary for building the Data Mining model, final testing and deployment tasks. The data manager layer will aid in dividing the data into multiple set so that it can be utilized during various stages of the data mining task. Same is the case with results of the Data Mining task, which might be utilized for further processing.
- **Input Data Flow :** The data need to be extracted from the database in the required format of the data mining task. Also the data flow needs to be controlled as per the Data Mining task requirements *i.e.,* row by row or bulk load. The Data Mining task may also require data in specific format (like itemized data for Associations). A few transformation routines will be necessary to transform the data from Database tier into the required format as per the specifications. Another option of transforming the data at database can be considered.
- **Output Data Flow:** The results generated by the data mining task will need to be managed and facilitated to target systems (Front End or other systems like CRM) in required data format and data flow specifications.

The Data Manager layer needs to be portable depending on the database from which data has to be extracted and the Data Mining tool.

2. Data Mining Tools/Algorithms

This is the heart of the complete architecture. The Data Mining tool perform different tasks. The prime functionality of the task is analyzing the data and generate the results. Various techniques / algorithms can be utilized depending upon the business case. These are described in the topic "data mining techniques".

Numerous tools are available in the market to give best possible result as output *e.g.,* SAS, SPSS, Teradata Miner and IBM Intelligent Miner. These tools merely facilitate the application of algorithms on the input data. But the most important task, which is always aligned to the specific business case, is setting the parameters for the algorithms.

3. Front End

The third tier is the Front-End layer, which facilitates the parameter settings for Data Mining application and visualization of the results in interpretable form.

Front End is the user interface layer. It has following prime functionalities.

- Administration
- Input Parameter Settings
- Data Mining Results / Visualization

1. Administration

Administration screens for the ETL and Data Mining tasks are usually provided as a part of the products/tools. These are utilized to administer the following primary tasks:

- Data flow processes (*e.g.,* Extracts, Loads).
- Data Mining routines.
- Error reporting and correction is also handled through the administration screens.
- User security settings.

2. Input Parameter Settings

During the building of Data Mining Model, iterations are inevitable. These iterations are needed to fine-tune the model by changing various parameters involved in the model. For executing a Data Mining task, the user needs to provide respective input parameters. Then observe the effect on the results and change the parameters if needed based on the interpretation and understanding of the results. This facility is provided in the Front End.

3. Data Mining Results

The results of the data mining task need formatting, conversion to user understandable form and reporting to the user. The front-end caters to the predefined formats of the out files generated by the respective Data Mining technique. The user will have the flexibility to view and analyze the results of Data Mining. Reporting utility performs the job of displaying the report, charts and smart reports (*e.g.*, Clusters, Trees, and Networks).

6.7. THE SCOPE OF DATA MINING

Now a day's almost every company including retail, finance, heath care, manufacturing transportation, and aerospace - are using data mining tools and techniques to take advantage of historical data. By using pattern recognition technologies and statistical and mathematical techniques through warehoused information, data mining helps analysts to recognize significant facts, relationships, trends, patterns, exceptions and anomalies that might otherwise go unnoticed.

Data mining technology can generate new business opportunities by:

6.7.1. Automated Prediction of Trends and Behaviours

Data mining automates the process of finding predictive information in large databases. Questions that traditionally required extensive hands-on analysis can now be answered directly from the data—quickly. A typical example of a predictive problem is targeted marketing. Data mining uses data on past promotional mailings to identify the targets most likely to maximize return on investment in future mailings. Other predictive problems include forecasting bankruptcy and other forms of default, and identifying segments of a population likely to respond similarly to given events.

6.7.2. Automated Discovery of Previously Unknown Patterns

Data mining tools sweep through databases and identify previously hidden patterns in one step. An example of pattern discovery is the analysis of retail sales data to identify seemingly unrelated products that are often purchased together. Other pattern discovery problems include detecting fraudul credit card transactions and identifying anomalous data that could represent data entry keying errors.

Data mining techniques can yield the benefits of automation on existing software and hardware platforms, and can be implemented on new systems as existing platforms are upgraded and new products developed. When data mining tools are implemented on high performance parallel processing systems, they can analyze massive databases in minutes. Faster processing means that users can automatically experiment with more models to understand complex data. High speed makes it practical for users to analyze huge quantities of data. Larger databases, in turn, yield improved predictions.

Databases can be larger in both depth and breadth.

More columns: Analysts must often limit the number of variables they examine when doing hands-on analysis due to time constraints. Yet variables that are discarded because they seem unimportant may carry information about unknown patterns. High performance data mining allows users to explore the full depth of a database, without pre-selecting a subset of variables.

More rows: Larger samples yield lower estimation errors and variance, and allow users to make inferences about small but important segments of a population.

6.8. DATA MINING TECHNIQUES [MDU May 2012]

The most commonly used techniques in data mining are:

- **Artificial Neural Networks:** These are Non-linear predictive models that learn through training and resemble biological neural networks in structure. Neural networks are an approach to computing that involves developing mathematical structures with the ability to learn. The methods are the result of academic investigations to model nervous system learning. Neural networks have the remarkable ability to derive meaning from complicated or imprecise data and can be used to extract patterns and detect trends that are too complex to be noticed by either humans or other computer techniques. A trained neural network can be thought of as an "expert" in the category of information it has been given to analyze. This expert can then be used to provide projections given new situations of interest and answer "what if" questions.

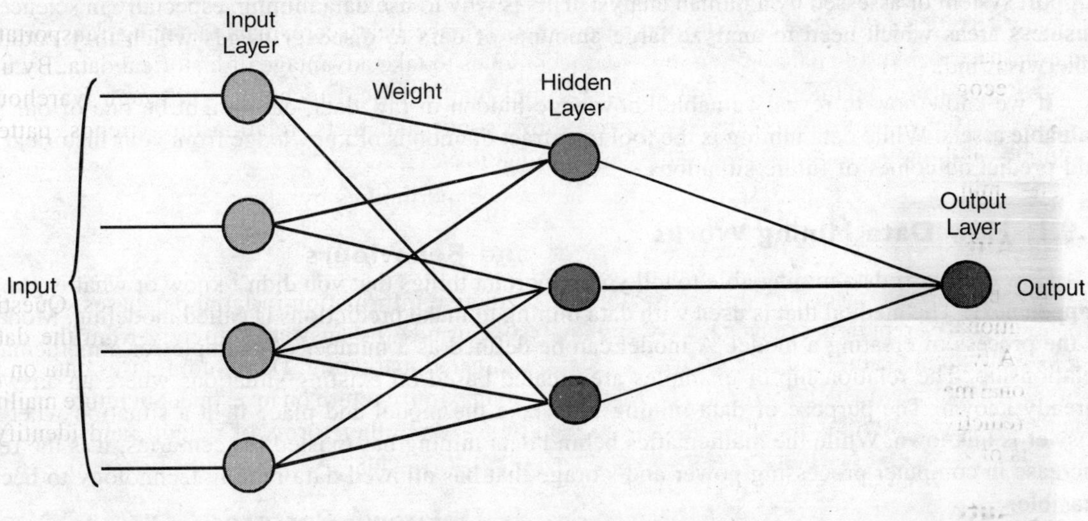

Fig. 6.7. Artificial Neural Network.

- **Decision Trees:** Decision trees are tree-shaped structures that represent sets of decisions. Decision trees are simple knowledge representation and they classify examples to a finite number of classes, the nodes are labelled with attribute names, the edges are labelled with possible values for this attribute and the leaves labelled with different classes. Objects are classified by following a path down the tree, by taking the edges, corresponding to the values of the attributes in an object these decisions generate rules for the classification of a dataset. Specific decision tree methods include Classification and Regression Trees (CART) and Chi Square Automatic Interaction Detection (CHAID).
- **Genetic Algorithms:** These are the Optimization techniques that use processes such as genetic combination, mutation, and natural selection in a design based on the concepts of evolution.
- **Nearest Neighbour Method:** A technique that classifies each record in a dataset based on a combination of the classes of the *k* record(s) most similar to it in a historical dataset (where *k* is greater than or equal to 1). Sometimes called the k-nearest neighbour technique.
- **Rule Induction:** Rule extraction is the extraction of useful if-then rules from data based on statistical significance.

Many of these technologies have been in use for more than a decade in specialized analysis tools that work with relatively small volumes of data. These capabilities are now evolving to integrate directly with industry-standard data warehouse and OLAP platforms. The data mining techniques will be explained in detail in coming chapters.

6.9. WHY USE DATA MINING?

Two main reasons to use data mining:

- Too much data and too little information
- There is a need to extract useful information from the data and to interpret the data.
- An increasingly competitive business environment drives the need for data mining.

Since enormous volumes of data are available and human analysts with no special tools is just wastage of time, money and other useful resources. And Data mining can automate the process of finding relationships and patterns in raw data and the results can be either utilized in an automated decision support system or assessed by a human analyst. This is why to use data mining, especially in science and business areas which need to analyze large amounts of data to discover trends which they could not otherwise find.

If we know how to reveal valuable knowledge hidden in raw data, data might be one of our most valuable assets. While data mining is the tool to extract diamonds of knowledge from your historical data and predict outcomes of future situations

6.9.1. How Data Mining Works

How exactly is data mining able to tell you important things that you didn't know or what is going to happen next? The method that is used with data mining to make predictions is called modeling. Modeling is the process of creating a model. A model can be defined as a number of examples or a mathematical relationship. The relationship or examples are created based on existing situations where an answer is already known. The purpose of data mining is to take the model and place it in a situation where the answer is unknown. While the mathematics behind data mining has existed for centuries, it is the recent increase in computer processing power and storage that has allowed data mining technology to become feasible.

Example: As a simple example of building a model, consider the director of marketing for a telecommunications company. He would like to focus his marketing and sales efforts on segments of the population most likely to become big users of long distance services. He knows a lot about his customers, but it is impossible to discern the common characteristics of his best customers because there are so many variables. From his existing database of customers, which contains information such as age, sex, credit history, income, zip code, occupation, etc., he can use data mining tools, such as neural networks, to identify the characteristics of those customers who make lots of long distance calls. For instance, he might learn that his best customers are unmarried females between the age of 30 and 40 who make in excess of $60,000 per year. This, then, is his model for high value customers, and he would budget his marketing efforts to accordingly.

6.10. DATA MINING TOOLS COMPARISON

Different data mining tools are available that work on different algorithms and on different platforms. Which tools is using which algorithm and working on which platform is shown in Fig. 6.8.

	Algorithms										Platforms		
Tool	NN	Tree	Naïve Bayes	k-Mns	k-NN	Stats	Pred	Time Series	Clust	Assoc	Win 32	UNIX	Par
IBM – Intelligent Miner	√	√			√		√	√	√	√	√	√	√
SPSS – Clementine	√	√					√	√	√	√	√	√	
SAS – Enterprise Miner	√	√				√	√	√	√	√	√	√	
SGI – MineSet		√	√	√			√		√	√	√	√	√
NeoVista Software – Decision Series	√	√	√				√		√	√	√	√	√

Fig. 6.8. Data Mining Tools Comparisons

where

NN = Neural Net

Tree = Decision Tree

k-Mns = K Means

k-NN = k-Nearest Neighbour

Stats = Traditional Statistical Techniques

Pred = Prediction

Clust = Clustering

Assoc = Association

Win 32 = Windows 95/98/NT

Par = Parallel Scalability (in at least one OS)

Brief description of all these data mining tools is given here.

1. IBM–Intelligent Miner

Operating environments: IBM AIX, AS/400, OS/390, and MVS/ESA servers; and Windows and NT clients.

Databases supported: DB/2 tables or flat files; through Intelligent Miner facilities, users can extract data from Oracle and Sybase.

Algorithms supported: Neural networks, decision trees, predictive modeling, association discovery, advanced statistics, logistics and linear regression, sequential pattern discovery and database segmentation.

Strengths

1. Intelligent Miner offers a complete set of datamining algorithms with a high degree of scalability and parallelism.
2. The products integrates with DB2 well, and provides various data transformations which other tools do not support for example the fields to records transformation.
3. All tools' report results in a visual format, and ready to present. Although the tool does not have OLAP capabilities, strong visualization compensates in part for that lack.

4. Miner supports three customizable applications:
   - **Customer Segmentation:** Segments and scores customer data from marketing databases for target marketing, cross-selling and customer retention campaigns etc.
   - **Market Basket Analysis:** Good for detecting/analyzing customer buying behaviour and predicting future behaviour by identifying affinities among customers' choices of products and services.
   - **Fraud Detection:** Identifies deviations from established norms in order to flag suspicious transactions indicative of fraudulent activity.
5. Multiple data mining steps can be combined into projects for unsupervised execution.

Weaknesses

1. The product runs only with db2 or flat files, and therefore a complicated extraction procedure needs to be followed to extract data from the oracle source. The tool does not perform data preparation (such as sampling) on textual data, but insists that the data be stored in db2. Moreover, intermediate results have to be stored in db2 files, with each step of the mining project stored independently (although combination into projects is possible).
2. Intelligent Miner lacks 3-D visualization capabilities.
3. The software lacks good documentation. The HTML documentation is obscure, and difficult to understand.
4. The only rule generator available in the product is tree based, and there is no support for non-hierarchical rule generation.

2. SPSS–Clementine

Operating Environments: Windows NT, HP-UX 9.X/10.X, IRIX 5.X/6.X, Solaris 2.X, Digital UNIX, AIX 4.2, DGUX 4.2, NCR UNIX SVR5 MP-RAS version 3.0 or higher.

Databases supported: All ODBC compliant databases, flat files and Clementine native files.

Algorithms supported: Neural Nets, Rule Induction, Association Rules (GRI, Apriori), C5.0 Rule Induction, Data Visualization, Kohonen Nets & Kmeans.

Strengths

1. Clementine supports all kinds of data. It can access data from an ASCII file, a database with ODBC support or a user specification. It has efficient sampling, sorting and selecting capabilities. Also, the results from data mining operations can be written back directly to ODBC databases.
2. Clementine has properly documented online help as well as reference manuals.
3. Clementine features a very nice end-user visual-programming environment for wiring together icons representing application functionality. It uses drag-and-drop icon development. Multiple data mining steps can be combined into projects for unsupervised execution.
4. Visualization is good for categorical variables and that can be used for finding associations.

Weaknesses

1. Clementine doesn't have many statistical functions to be performed on the data. The only statistical function that it provides is linear regression.
2. Clementine doesn't work well with noisy data. The quality of data needs to be quite high for any significant results.
3. Clementine finds associations best on column-wise data. While performing market-basket analysis, if the data are not in such format, it needs to be transformed which requires a lot of effort. Also the data type should be categorical.
4. Clementine doesn't support multi-threading *i.e.,*, no two processes can be simultaneously run.

3. SAS Institute–Enterprise Miner

Operating environments: Windows NT, Sun Solaris, IBM AIX, and HP/UX servers; and Windows NT and 95 clients.

Databases supported: DB/2, Excel, SAS Data Store, Informix, Oracle, Sybase, and others.

Algorithms supported: Decision trees, neural networks, clustering, classification and regression trees, advanced statistics and market basket analysis.

Strengths

1. Supports several major algorithms so users don't have to mix and match software from different vendors. Strong technical support.
2. Enterprise miner features an icon-based GUI and uses SAS's SEMMA data mining methodology (Sample, Explore, Modify, Model and Assess).
3. It works fine as a general-purpose datamining tool as well as for CRM (Customer Relationship Management), credit-risk scoring, fraud detection, and market-basket analysis.
4. Enterprise Miner can operate as a standalone application or integrated with SAS dataware housing solutions. Eminer also has APIs to support the exchange of data with other vendor's packaged applications.

Weaknesses

1. It could not print a GUI-based graph comparing results from several different algorithms, including logistical regression, decision tree, and neural network.
2. Enterprise Miner lacks scalability and the ability to handle large datasets.

4. Silicon Graphics–MineSet

Operating environments: SGI Irix 6.0 and higher, Windows NT, Windows 95/98 for client only.

Databases supported: Direct connection to Informix, Oracle, and Sybase and connection to other RDBMS through ODBC and flat files.

Algorithms supported: Association rules, decision trees, evidence trees, option trees, regression, and clustering.

Strengths

1. Offers a wide variety of visualization techniques, including map and scatter plots, as well as the ability to create customized visuals.
2. Mine set has data analysis tools like :
   - ➡ Association Rule Generator - analyzes data to reveal the nature and frequency or relationships between data types
   - ➡ Rule Visualizer - graphically displays results from rule generator
   - ➡ Tree Visualizer - displays 3-D graphics for visualizing hierarchically structured data in a fly-through landscape format.
   - ➡ Map Visualizer – explores data with geographical relationships for trend and pattern identification
   - ➡ Scatter Visualizer – helps analyze behaviour of data in many dimensions at once.

   Data integration tools provide user with complete control over mapping of data to analysis tools.
3. Mineset provides APIs to support the exchange of data with other vendor's packaged applications. It also provides plug-ins for adding new functionality.

Weaknesses

1. Mineset has only a few statistical techniques while other packages have more robust statistical techniques.
2. Mineset requires extensive data preparation for market-basket analysis kind of applications as it can work only with column-wise data.
3. It needs an industry-standard API, which the company is developing to support easier development of vertical applications.
4. Mineset was built to present three-dimensional scatter plots, but for a simple two-dimensional scatter plot where the x-axis starts in the bottom left-hand corner, the numbers show up backwards or not at all.

5. NeoVista Software–Decision Series

Operating environments: HP/UX, Sun Solaris, Digital UNIX servers; and Windows NT and 95 clients.

Databases supported: ODBC-compliant databases and native connections to Oracle and Informix.

Algorithms supported: Neural networks, decision networks, decision trees, decision clusters, and association rules.

Strengths

1. Decision Series is a high-end, enterprise-level set of tools built using object-oriented technology and designed to work in parallel, scalable environments.
2. Highly scalable for parallel environments. Strong services associated with the software.

Weaknesses

1. Runs only on UNIX servers.
2. Successful use requires a statistician with quantitative analysis experience.

6.11. ADVANTAGES AND DISADVANTAGES OF DATA MINING

Data mining is an important part of knowledge discovery process that analyzes large enormous set of data and gives us unknown, hidden and useful information and knowledge. Data mining has not only applied effectively in business environment but also in other fields such as weather forecast, medicine, transportation, healthcare, insurance, government and etc. Data mining brings a lot of advantages when using in a specific industry. Beside those advantages, data mining also has its own disadvantages as well such as privacy, security and misuse of information. We will examine the advantage and disadvantages of

6.11.1. Advantages of Data Mining

1. Marketing/Retail

Data mining helps marketing companies to build models based on historical data to predict who will respond to new marketing campaign such as direct mail, online marketing campaign and etc. Through this prediction, marketers can have appropriate approach to sell profitable products to targeted customers with high satisfaction.

Data mining brings a lot of benefits to retail company in the same way as marketing. Through market basket analysis, the store can have an appropriate production arrangement in the way that customers can buy frequent buying products together with pleasant. In addition, it also help the retail company offers a certain discount for particular products what will attract customers.

2. Finance/Banking

Data mining gives financial institutions information about loan information and credit reporting. By building a model from previous customer's data with common characteristics, the bank and financial can estimate what are the god and/or bad loans and its risk level. In addition, data mining can help banks to detect fraudulent credit card transaction to help credit card's owner prevent their losses.

3. Manufacturing

By applying data mining in operational engineering data, manufacturers can detect faulty equipments and determine optimal control parameters. For example semi-conductor manufacturers had a challenge that even the conditions of manufacturing environments at different wafer production plants are similar, the quality of wafer are lot the same and some for unknown reasons even contain defects. Data mining has been applied to determine the ranges of control parameters that lead to the production of golden wafer. Then those optimal control parameters are used to manufacture wafers with desired quality.

4. Governments

Data mining helps government agency by digging and analyzing records of financial transaction to build patterns that can detect money laundering or criminal activity.

6.11.2. Disadvantages of Data Mining

1. Privacy Issues

The concerns about the personal privacy have been increasing enormously recently especially when internet is booming with social networks, e-commerce, forums, blogs.... Because of privacy issues, people are afraid of their personal information is collected and used in unethical way that potentially causing them a lot of trouble. Businesses collect information about their customers in many ways for understanding their purchasing behaviours trends. However businesses don't last forever, some days they may be acquired by other or gone. At this time the personal information they own probably is sold to other or leak.

2. Security Issues

Security is a big issue. Businesses owns information about their employee and customers including social security number, birthday, payroll and etc. However how properly this information is taken is still in questions. There have been a lot of cases that hackers were accesses and stole big data of customers from big corporation such as Ford Motor Credit Company, Sony... with so much personal and financial information available, the credit card stolen and identity theft become a big problem.

3. Misuse of Information/Inaccurate Information

Information collected through data mining intended for marketing or ethical purposes can be misused. This information is exploited by unethical people or business to take benefit of vulnerable people or discriminate against a group of people.

In addition, data mining technique is not perfectly accurate therefore if inaccurate information is used for decision-making will cause serious consequence.

6.12. APPLICATIONS OF DATA MINING

Data mining is a process that analyzes the large amount of data to find the new and hidden information that improves business efficiency. Various industries have been adopt data mining to their mission-critical business processes to gain competitive advantages and help business grows. This tutorial illustrates some data mining applications in sale/marketing, banking/finance, health care and insurance, transportation and medicine.

1. Data Mining Applications in Sales/Marketing

Data mining enables the businesses to understand the patterns hidden inside past purchase transactions, thus helping in plan and launch new marketing campaigns in prompt and cost effective way. The following illustrates several data mining applications in sale and marketing.

- Data mining is used for market basket analysis to provides insight information on what product combinations were purchased, when they were bought and in what sequence by customers. This information helps businesses to promote their most profitable products to maximize the profit. In addition, it encourages customers to purchase related products that they may have been missed or overlooked.
- Retails companies uses data mining to identify customer's behaviour buying patterns.

2. Data Mining Applications in Banking/Finance

- Several data mining techniques such as distributed data mining has been researched, modelled and developed to help credit card fraud detection.
- Data mining is used to identify customers loyalty by analyzing the data of customer's purchasing activities such as the data of frequency of purchase in a period of time, total monetary value of all purchases and when was the last purchase. After analyzing those dimensions, the relative measure is generated for each customer. The higher of the score, the more relative loyal the customer is.
- To help bank to retain credit card customers, data mining is used. By analyzing the past data, data mining can help banks to predict customers that likely to change their credit card affiliation so they can plan and launch different special offers to retain those customers.
- Credit card spending by customer groups can be identified by using data mining.
- The hidden correlation's between different financial indicators can be discovered by using data mining.
- From historical market data, data mining enable to identify stock trading rules.

3. Data Mining Applications in Health Care and Insurance

The growth of the insurance industry is entirely depends on the ability of converting data into the knowledge, information or intelligence about customers, competitors and its markets. Data mining is applied in insurance industry lately but brought tremendous competitive advantages to the companies who have implemented it successfully. The data mining applications in insurance industry are listed below:

- Data mining is applied in claims analysis such as identifying which medical procedures are claimed together.
- Data mining enables to forecasts which customers will potentially purchase new policies.
- Data mining allows insurance companies to detect risky customers' behaviour patterns.
- Data mining helps detect fraudulent behaviour.

4. Data Mining Applications in Transportation

Data mining helps to determine the distribution schedules among warehouses and outlets and analyze loading patterns.

5. Data Mining Applications in Medicine

- Data mining enables to characterize patient activities to see coming office visits.
- Data mining help identify the patterns of successful medical therapies for different illnesses.

Data mining applications are continuously developing in various industries to provide more hidden knowledge that enable to increase business efficiency and grow businesses.

6.13. KDD-DATA MINING ISSUES AND CHALLENGES

Data Mining applications rely on databases to supply the raw data for input. The issues in the databases / data (*e.g.,* volatility, incompleteness, noise, and volume) augment the issues by the time it reaches data mining task. Other problems arise as a result of the adequacy and relevance of the information stored.

1. Limited Information

A database is often designed for purposes different from data mining and sometimes the properties or attributes that would simplify the learning task are not present nor can they be requested from the real world. Inconclusive data causes problems because if some attributes essential to knowledge about the application domain are not present in the data it may be impossible to discover significant knowledge about a given domain. For example malaria can not be diagnosed from a patient database if that database does not contain the patient's red blood cell count.

2. Noisy and Missing Values

Databases are usually contaminated by errors so it cannot be assumed that the data they contain is entirely correct. Attributes which rely on subjective or measurement judgments can give rise to errors such that some examples may even be mis-classified. Errors in either the values of attributes or class information are known as noise. Obviously where possible it is desirable to eliminate noise from the classification information as this affects the overall accuracy of the generated rules.

Missing data can be treated by discovery systems in a number of ways such as;

- simply disregard missing values
- omit the corresponding records
- infer missing values from known values
- treat missing data as a special value to be included additionally in the attribute domain
- or average over the missing values using Bayesian techniques.

Noisy data in the sense of being imprecise is characteristic of all data collection and typically fit a regular statistical distribution such as Gaussian while wrong values are data entry errors. Statistical methods can treat problems of noisy data, and separate different types of noise. We will distinguish between two forms of noise in the data, both described below:

(a) **Corrupted Values:** Sometimes some of the values in the training set are altered from what they should have been. This may result in one or more tuples in the database conflict with the rules already established. The system may then regard these extreme values as noise, and ignore them. Alternatively, one may take the values into account possibly changing correct patterns recognized. The problem is that one never knows if the extreme values are correct or not, and the challenge is how to handle "weird" values in the best manner.

(b) **Missing Attribute Values:** One or more of the attribute values may be missing both for examples in the training set and for object which are to be classified. If attributes are missing in the training set, the system may either ignore this object totally, try to take it into account by for instance finding what is the missing attribute's most probable value, or use the value "unknown" as a separate value for the attribute. When an attribute value is missing for an object during classification, the system may check all matching rules and calculate the most probable classification.

3. Uncertainty

Uncertainty refers to the severity of the error and the degree of noise in the data. Data precision is an important consideration in a discovery system.

4. Size, Updates, and Irrelevant Fields

Databases tend to be large and dynamic in that their contents are ever-changing as information is added, modified or removed. The problem with this, from the data mining perspective is how to ensure that the rules are up-to-date and consistent with the most current information. Also the learning system has to be time-sensitive as some data values vary over time and the discovery system is affected by the 'timeliness' of the data. Another problem is difficult training set.

Sometimes the training set is not the ultimate training set due to several reasons. These are the following:

Not Representative Data: If the data in the training set is not representative for the objects in the domain, then there is a problem. If rules for diagnosing patients are being created and only elderly people are registered in the training set, the result for diagnosing a kid based on these data probably will not be good. Even though this may have serious consequences, we would say that not representative data is mainly a problem of machine learning when the learning is based on few examples. When using large data sets, the rules created probably are representative, as long as the data being classified belongs to the same domain as those in the training set.

No Boundary Cases: To find the real differences between two classes, some boundary cases should be present. If a data mining system for instance is to classify animals, the property counting for a bird might be that it has wings and not that it can fly. This kind of detailed distinction will only be possible if the birds are registered *e.g.,* penguins are registered.

Limited Information: In order to classify an object to a specific class, some condition attributes are investigated. Sometimes, two objects with the same values for condition attributes have a different classification. Then, the objects have some properties which are not among the attributes in the training set, but still make a difference. This is a problem for the system, which does not have any way of distinguish these two types of objects.

5. Databases are Dynamic

Databases usually change continually. We would like rules which reflect the content of the database at all times, in order to make the best possible classification. Many existing data mining systems require that all the training examples are given at once. If something is changed at a later time, the whole learning process may have to be conducted again. An important challenge for data mining systems is to avoid this, and instead change its current rules according to updates performed

6. Databases May be Huge

The size of databases seem to be ever increasing. Most machine learning algorithms have been created for handling only a small training set, for instance a few hundred examples. In order to use similar techniques in databases thousands of times bigger, much care must be taken. Having very much data is advantageous since they probably will show relations really existing, but the number of possible descriptions of such a dataset is enormous. Some possible ways of coping with this problem, are to design algorithms with lower complexity and to use heuristics to find the best classification rules. Simply using a faster computer is seldom a good solution.

7. Scalability

To effectively extract the information from a huge amount of data in the database, data mining algorithms must be efficient and scalable. Efficiency means that the running time of an algorithm must be pre-dictable and acceptable.

Summary

- **KDD** refers to the overall process of discovering useful knowledge from data. It involves the evaluation and possibly interpretation of the patterns to make the decision of what qualifies as knowledge. It also includes the choice of encoding schemes, preprocessing, sampling, and projections of the data prior to the data mining step.
- **Data mining** refers to the application of algorithms for extracting patterns from data without the additional steps of the KDD process.
- Knowledge discovery in databases is the non-trivial process of identifying valid, novel, potentially useful, and ultimately understandable patterns in data.

Data	A set of facts, F.
Pattern	An expression E in a language L describing facts in a subset F_E of F.
Process	KDD is a *multi-step process* involving data preparation, pattern searching, knowledge evaluation, and refinement with iteration after modification.
Valid	Discovered patterns should be true on new data with some degree of certainty. Generalize to the future (other data).
Novel	Patterns must be novel (should not be previously known).
Useful	Actionable; patterns should potentially lead to some useful actions.
Understandable	The process should lead to human insight. Patterns must be made understandable in order to facilitate a better understanding of the underlying

- The Primary Tasks of Data Mining are prediction and description and both cane be achieved by classification, regression, clustering summarization, dependency modelling etc.
- Popular Data Mining Methods are
  (*a*) Decision Trees and Rules
  (*b*) Non-linear Regression and Classification Methods
  (*c*) Example-based Methods
  (*d*) Probabilistic Graphical Dependency Models
  (*e*) Relational Learning Models
- **Cycle of data mining**

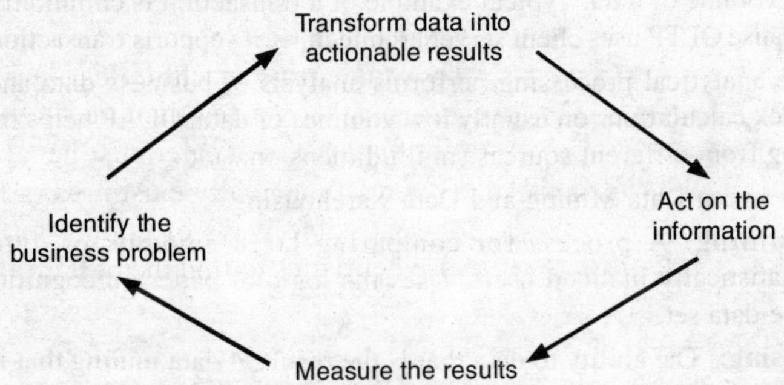

- Data mining systems can be classified according to
  (*a*) *Kinds of knowledge mined:* That is based on data mining functionalities like association, classification, clustering, evolution analysis etc.

(*b*) *Kinds of database mined:* that is according to the data models like, relational, transactional, object-oriented etc

(*c*) *Kinds of techniques used;* that is according to the data analysis methods are employed like visualization, neural networks, Artificial intelligence etc.

(*d*) *Application adapted:* the data mining systems specially meant for finance, healthcare, telecommunication etc.

- **Challenges in Data mining:**
  - Developing a unifying theory of data mining
  - Scaling up for high dimensional data and high speed data streams
  - Mining sequence data and time series data
  - Mining complex knowledge from complex data
  - Data mining in a network setting
  - Distributed data mining and mining multi-agent data
  - Data mining for biological and environmental problems
  - Data Mining process-related problems
  - Security, privacy and data integrity
  - Dealing with non-static, unbalanced and cost-sensitive data

Review Questions

1. Differentiate between data mining and database management systems.
2. Is there any difference between KDD and Data mining.
3. What do you mean by data purging?

   Hint: The process of cleaning junk data is termed as data purging. Purging data would mean getting rid of unnecessary NULL values of column
4. What are the different problems that "Data mining" can solve?
5. Explain different stages of data mining.
6. What is Discrete and Continuous data in Data mining world?
7. What is MODEL in Data mining world?
8. What is decision tree algorithm
9. What are OLAP and OLTP?

   Hint: OLTP: Online Transaction and Processing helps and manages applications based on transactions involving high volume of data. Typical example of a transaction is commonly observed in Banks, Air tickets etc. Because OLTP uses client server architecture, it supports transactions to run cross a network.

   OLAP: Online analytical processing performs analysis of business data and provides the ability to perform complex calculations on usually low volumes of data. OLAP helps the user gain an insight on the data coming from different sources (multi dimensional).
10. Differentiate between Data Mining and Data warehousing.

    Hint: Data mining: A process for comparing large amounts of data for finding patterns. Intended for statistically inclined users. Uses the logic of pattern recognition for identifying trends within a sample data set.

    Data warehousing: The ability to data that is the result of data mining that is to be used in future for database inquiries. Intended for the users those are to be data experts who analyze by business dimensions directly. A process used for extracting and storing data which allows easier reporting.

❑❑❑

7 Chapter

Data Mining Query Language

INSIDE THIS CHAPTER

7.1. Introduction **7.2.** Importance of Data Mining Query Language **7.3.** Syntax for Task-Relevant Data Specifications **7.4.** DMQL: Syntax For Specification of Task-Relevant Data **7.5.** Syntax for Specifying the Kind of Knowledge to be Mined **7.6.** To Specify What Concept Hierarchies to Use **7.7.** Syntax for Interestingness Measure Specification **7.8.** Syntax for Pattern Presentation and Visualization Specification **7.9.** Architecture of DM Systems

In this chapter data mining query language which provides a foundation on which user-friendly interfaces can be built is discussed.

7.1. INTRODUCTION

Data Mining is the process of discovering interesting knowledge (patterns) from large amounts of data stored either in databases, data warehouses, or other information repositories. But there is a misconception that DM systems can automatically dig out all the information without human intervention. But the human intervention is required because if this is not present then system would uncover a large set of patterns that may exceed the size of the database. Therefore the communication of the system with the user is provided by data mining primitives.

These primitives allow the user to *interactively* communicate with the data mining system during discovery in order to direct the mining process, or examine the findings from different angles or depths.

1. The set of *task-relevant data* to be mined: This specifies the portions of the database or the set of data in which the user is interested. This includes the database attributes or data warehouse dimensions of interest (referred to as the *relevant attributes or dimensions*
2. The *kind of knowledge* to be mined: This specifies the *data mining functions* to be performed, such as characterization, discrimination, association or correlation analysis, classification, prediction, clustering, outlier analysis, or evolution analysis.
3. The *background knowledge* to be used in the discovery process: This knowledge about the domain to be mined is useful for guiding the knowledge discovery process and for evaluating the patterns

found. *Concept hierarchies* are a popular form of background knowledge, which allow data to be mined at multiple levels of abstraction.

4. The *interestingness measures and thresholds* for pattern evaluation. They may be used to guide the mining process or, after discovery, to evaluate the discovered patterns.

   Different kinds of knowledge may have different interestingness measures. For example, interestingness measures for association rules include *support* and *confidence*. Rules whose support and confidence values are below user-specified thresholds are considered uninteresting.

5. The expected *representation for visualizing* the discovered patterns: This refers to the form in which discovered patterns are to be displayed, which may include rules, tables, charts, graphs, decision trees, and cubes.

   A data mining query language can be designed to incorporate these primitives, allowing users to flexibly interact with data mining systems. Having a data mining query language provides a foundation on which user-friendly graphical interfaces can be built.

This facilitates a data mining system's communication with other information systems and its integration with the overall information processing environment. Designing a comprehensive data mining language is challenging because data mining covers a wide spectrum of tasks, from data characterization to evolution analysis. Each task has different requirements. The design of an effective data mining query language requires a deep understanding of the power, limitation, and underlying mechanisms of the various kinds of data mining tasks.

7.2. IMPORTANCE OF DATA MINING QUERY LANGUAGE

Some of the importance of data mining query language are:

1. DMQL allows the ad hoc mining of several kinds of knowledge from relational databases and data warehouses at multiple levels of abstraction.
2. Having a good query language for data mining may help standardize the development of platforms for data mining systems.
3. By providing a standardized language like SQL
   - There is a hope to achieve a similar effect like that SQL has on relational database
   - It provides a foundation for system development and evolution
   - It will facilitate information exchange, technology transfer, commercialization and wide acceptance.

7.3. SYNTAX FOR TASK-RELEVANT DATA SPECIFICATIONS

Task related data: The first step in defining a data-mining task is the specification of the task-relevant data, that is, the data on which mining is to be performed. This involves specifying

1. Database or data warehouse name
2. Database tables or data warehouse cubes.
3. Condition for data selection
4. Relevant attributes or dimensions
5. Data grouping criteria

1. Syntax for Task-Relevant Data Specification

use database database_name, or *use data warehouse* data_warehouse_name

- *from relation*(s)/cube(s) [*where* condition]
- *in relevance to* att_or_dim_list

- *order by* order_list
- *group by* grouping_list
- *having* condition

Example: This example shows how to use DMQL to specify the task-relevant data, the mining of associations between items frequently purchased at AllElectronics by Australian customers, can be specified by providing the following information.

- Name of the *database or data warehouse* to be used (*e.g., AllElectronics_db*).
- Names of the *tables or data cubes* containing relevant data (*e.g., item, customer, purchases* and *items_sold*).
- *Conditions* for selecting the relevant data (*e.g.,* retrieve data pertaining to purchases made in Canada for the current year).
- The *relevant attributes or dimensions* (*e.g., name* and *price* from the *item* table and income and age from the customer table)

2. Types of Knowledge to be Mined

It is important to specify the knowledge to be mined as this determines the data mining function to be performed. The kinds of knowledge include

- Characterization Association
- Classification/prediction
- Clustering
- Outlier analysis

and so on.

User can also provide the pattern templates also called metarules, metaqueries or metapattern.

3. Background Knowledge

It is the information about the domain to be mined. **Concept hierarchies is a powerful form of background knowledge.** Concept hierarchy defines the sequence of mapping from low level abstraction to high level abstraction. This means that data can be mined at multiple level of abstraction. The concept hierarchy allow the users to view data from different perspectives, so that the relationship will be more clear between the concepts. **Example :** location. There may be more than one concept hierarchy for a given attribute or dimension based on different user viewpoints.

Example: Regional sales manager may prefer the previous concept hierarchy but marketing manager might prefer to see location with respect to linguistic lines in order to facilitate the distribution of commercial ads.

Four major types of concept hierarchy are:

- **Schema hierarchy :** Schema hierarchy is the total or partial order among the attributes in the database schema. Schema hierarchy provides metadata information.

  Example : street < city < province_or_state < country.
- **Set-grouping hierarchy :** Set grouping hierarchies organizes the value for a given attribute into groups or sets or range of values. The total as well as partial ordering can be defined among the sets using Set grouping hierarchy. The importance of Set grouping hierarchies is that they are used to enrich the schema defined hierarchy.
- Example: Set-grouping hierarchy for age

  {young, middle_aged, senior} ⊂ all (age)

  {20....29} ⊂ young

{40....59} ⊂ middle_aged

{60....89} ⊂ senior

- **Operation-derived hierarchy :** Operation derived hierarchy is based on the operations. Specified operations may include the decoding of an encoded strings, extraction of information from complex data structures, clustering the objects.

  Example : e-mail address, login-name < department < university < country
- **Rule-based hierarchy :** Rule based hierarchy occurs when either whole or portion of concept hierarchy is defined as set of rules and is evaluated based on current database data and rule definition.

  Example : Following rules are used to categorize items as *low_profit, medium_profit* and *high_profit_margin.*

  low_profit_margin(X) <= price(X, P1)^cost(X,P2)^((P1-P2)<50)

  medium_profit_margin(X) <= price (X,P1)^cos(X,P2)^((P1-P2)≥50)^((P1-P2)≤250)

  high_profit_margin(X) <= price(X,P1)^cost(X,P2)^((P1-P2)>250)
- **Pattern Interestingness Measurements:** The pattern interestingness measurement are
  - **Simplicity:**

    Example: rule length.
  - **Certainty (confidence):** Assesses the validity or trustworthiness of a pattern. confidence is a certainty measure

    $$confidence\ (A{=}{>}B) = \frac{\#\,tuples\ containing\ both\ A\ and\ B}{\#\ tuples\ containing\ A}$$

    A confidence of 85% for the rule buys(X, "computer")=>buys(X,"software") means that 85% of all customers who purchased a computer also bought software.
  - **Utility (support):** Useful of a pattern

    $$support\ (A{=}{>}B) = \frac{\#\,tuples\ containing\ both\ A\ and\ B}{total\ \#\ of\ tuples}$$

    A support of 30% for the previous rule means that 30% of all customers in the computer department purchased both a computer and software.
  - Association rules that satisfy both the minimum confidence and support threshold are referred to as **strong association rules.**
  - **Novelty:** Patterns contributing new information to the given pattern set are called novel patterns (example: Data execution).

    removing redundant patterns is a strategy for detecting novelty.

Visualization of Discovered Patterns

For data mining to be effective the data mining system should be able to display the discovered patterns in the form of rules, tables, crosstabs, pie/bar chart, etc. but the mode of presentation should be specified by the user because different knowledge requires different representation.

7.4. DMQL: SYNTAX FOR SPECIFICATION OF TASK-RELEVANT DATA

The name of the relevant database or datawarehouse , conditions, attributes and dimensions must be specified.

- use database **<database_name>**

  or use data warehouse **<data_warehouse_name >**

- from **<relation(s)/cube(s)>**[where **condition**]
  - **in relevance to** <attribute_or_dimension_list>
  - **order by** <order_list>
  - **group by** <grouping_list>
  - **having** <condition>

Example: This example shows how to use DMQL to specify the task-relevant data described for the mining of associations between items frequently purchased at *AllElectronics* by Canadian customers, with respect to customer *income* and *age.* In addition, the user specifies that she would like the data to be grouped by date. The data are retrieved from a relational database.

Use database AllElectronics_db

in relevance to I.name, I.price, C.income, C.age

from customer C, item I, purchases P, items_sold S

where I.item_ID = S.item_ID and S.trans_ID = P.trans_ID and P.cust_ID = C.cust_ID and C.address = "Canada"

group by P.date.

7.5. SYNTAX FOR SPECIFYING THE KIND OF KNOWLEDGE TO BE MINED

- **Characterization:**

```
<Mine_Knowledge_Specification> ::=
    mine characteristics [as <pattern_name>]
    analyze <measure(s)>
```

Example:

```
mine characteristics as customerPurchasting analyze count%
```

- **Discrimination:**

```
<Mine_Knowledge_Specification> ::=
    mine comparison [as <pattern_name>]
    for <target_class> where <target_condition>
    {versus <contrast_class_i where <contrast_condition_i>}
    analyze <measure(s)>
```

Example:

```
Mine comparison as purchaseGroups
    for bigspenders where avg(I.price) >=$100
    versus budgetspenders where avg(I.price) < $100
    analyze count
```

- **Association**

```
mine associations [as pattern_name]
```

- **Classification**

```
mine classification [as pattern_name] analyze
    classifying_attribute_or_dimension
```

- **Prediction**

```
mine prediction [as pattern_name] analyze
    prediction_attribute_or_dimension {set {attribute_or_dimension_i=
    value_i}}
```

- **Association:**

```
<Mine_Knowledge_Specification> ::=
    mine comparison [as <pattern_name>]
    [matching <metapattern>]
```

Example:

```
mine associations as buyingHabits
   matching P(X: customer, W) ^ Q(X, Y) => buys(X,Z)
```

➡ **Classification:**

```
<Mine_Knowledge_Specification> ::=
   mine classification [as<pattern_name>]
   analyze <classifying_attribute_or_dimension>
```

Example:

```
mine classification as classifyCustomerCreditRating
   analyze credit_rating
```

7.6. SYNTAX FOR CONCEPT HIERARCHY SPECIFICATION

To Specify What Concept Hierarchies to Use

```
use hierarchy <hierarchy name >for <attribute_or_dimension>
```

We use different syntax to define different type of hierarchies

➡ schema hierarchies (in this the order is more important)

```
define hierarchy time_hierarchy on date as [date,month quarter,year]
```

➡ set-grouping hierarchies

```
define hierarchy age_hierarchy for age on customer as
level1: {young, middle_aged, senior} < level0: all
level2: {20, ..., 39} < level1: young
level2: {40, ..., 59} < level1: middle_aged
level2: {60, ..., 89} < level1: senior
```

➡ operation-derived hierarchies

```
define hierarchy age_hierarchy for age on customer as
{age_category(1), ..., age_category(5)} := cluster(default, age, 5)
< all(age)
```

➡ rule-based hierarchies

```
define hierarchy profit_margin_hierarchy on item as

level_1: low_profit_margin < level_0: all
if (price - cost) ? $50
level_1: medium-profit_margin < level_0: all
if ((price - cost) > $50) and ((price - cost) ? $250))
level_1: high_profit_margin < level_0: all
if (price - cost) > $250
```

7.7. SYNTAX FOR INTERESTINGNESS MEASURE SPECIFICATION

➡ Interestingness measures and thresholds can be specified by the user with the statement:

```
with <interest_measure_name> threshold = threshold_value
```

Example:

```
with support threshold = 0.05
with confidence threshold = 0.7??
```

7.8. SYNTAX FOR PATTERN PRESENTATION AND VISUALIZATION SPECIFICATION

[MDU Dec. 2011]

We have syntax which allows users to specify the display of discovered patterns in one or more forms.

```
display as <result_form>
```

To facilitate interactive viewing at different concept levels, the following syntax is defined:

```
<Multilevel_Manipulation> ::=
roll up on <attribute_or_dimension >
drill down on <attribute_or_dimension >
add <attribute_or_dimension>
drop <attribute_or_dimension>
```

Putting It All Together: the Full Specification of a DMQL Query

```
use database OurVideoStore_db
use hierarchy location_hierarchy for B.address
mine characteristics as customerRenting
analyze count%
in relevance to C.age, I.type, I.place_made
from customer C, item I, rentals R, items_rent S, works_at W, branch
where I.item_ID = S.item_ID and S.trans_ID = R.trans_ID

and R.cust_ID = C.cust_ID and R.method_paid = "Visa"

and R.empl_ID = W.empl_ID and W.branch_ID = B.branch_ID and B.address
= "Alberta" and I.price >= 100

with noise threshold = 0.05
display as table
```

Example: Suppose that user would like to describe the general characteristics of graduate students in the Big-University database, given the attributes name, gender, major, birth_place, birth_date, residence, phone# (telephone number), and gpa (grade_point_average). A data mining query for this characterization can be expressed in the data mining query language DMQL as follows:

```
use Big_University_DB

mine characteristics as"Arts.Students"

in relevance to name, gender, major, birth_place, birth_date,
residence, phoned, gpa

from student where Status in "graduate"
```

7.9. ARCHITECTURE OF DM SYSTEMS

With the diverse application of DM, it is expected that a good variety of DM systems must be designed. Since we know that DM is the extraction of the useful information and the information is stored in the data warehouse.

Therefore the critical question arises that how to integrate the Data mining with Data warehouse. This gives rise to four architectures. These are:

(*i*) No coupling
(*ii*) Loose coupling
(*iii*) Semi-tight coupling
(*iv*) Tight coupling

No Coupling: DM system will not utilize any functionality of a DB or DW system.

Loose Coupling: DM system will use some facilities of DB and DW system like storing the data in either of DB or DW systems and using these systems to data retrieval.

Semi-tight Coupling: Besides linking a DM system to a DB/DW systems, efficient implementation of a few DM primitives.

Tight Coupling: DM system is smoothly integrated with DB/DW systems. Each of these DM, DB/DW is treated as main functional component of information retrieval system.

Other Data Mining Languages and Standardization Effects

Association rule language specifications:

➡ MSQL (Imielinski & Virmani'99)
➡ MineRule (Meo Psaila and Ceri'96)
➡ Query flocks based on Datalog syntax (Tsur et al'98)

OLEDB for DM (Microsoft'2000)

➡ Based on OLE, OLE DB, OLE DB for OLAP
➡ Integrating DBMS, data warehouse and data mining

CRISP-DM (CRoss-Industry Standard Process for Data Mining):

➡ Providing a platform and process structure for effective data mining
➡ Emphasizing on deploying data mining technology to solve business problems

Designing Graphical User Interfaces based on a data mining query language

What tasks should be considered in the design GUIs based on a data mining query language?

➡ Data collection and data mining query composition
➡ Presentation of discovered patterns
➡ Hierarchy specification and manipulation
➡ Manipulation of data mining primitives
➡ Interactive multilevel mining
➡ Other miscellaneous information

Summary

- Primitives for specifying a data mining task are:
  1. Task relevant data.
  2. Database or data warehouse name!
  3. Database tables or data warehouse cubes!
  4. Conditions for data selection!
  5. Relevant attributes or dimensions!
  6. Data grouping criteria
  7. Knowledge type to be mined!

8. Characterization!
9. Discrimination!
10. Association/correlation!
11. Classification/prediction!
12. Clustering
13. Background knowledge !
14. Concept hierarchies!
15. User beliefs about relationships in the data
16. Pattern interestingness measures!
17. Simplicity!
18. Certainty (e.g., confidence)!
19. Utility (e.g., support)!
20. Novelty
21. Visualization of discovered patterns!
22. Rules, tables, reports, charts, graphs, decision trees,! and cubes!
23. Drill-down and roll-up

- **DMQL**
  - ➡ A DMQL can provide the ability to support ad-hoc and interactive data mining.
  - ➡ By providing a standardized language like SQL, we hope to achieve the same effect that SQL have on relational database.

  Design: DMQL is designed with the primitives described earlier.

  Concept hierarchies is also important to:

  - ➡ discovered knowledge might be more understandable when represented at high concept level.
  - ➡ Interactive drill up/down, pivoting, slicing and dicing provide different perspective to data.
- **Data Mining System Architecture.**

  Coupling data mining system with DB/DW system

  - ➡ No coupling—flat file processing,
  - ➡ Loose coupling

  Fetching data from DB/DW

  - ➡ Semi-tight coupling—enhanced DM performance

  Provide efficient implement a few data mining primitives in a DB/DW system—sorting, indexing, aggregation, histogram analysis, multiway join, precomputation of some stat functions.

  Tight coupling—A uniform information processing environment

  - ➡ DM is smoothly integrated into a DB/DW system, mining query is optimized based on mining query, indexing, query processing methods.
- **Data Processing** is to get the required information from huge, incomplete, noisy and inconsistent set of data it is necessary to use dataprocessing.
- Data cleaning is a procedure to "clean" the data by filling in missing values, smoothing noisy data, identifying or removing outliers, and resolving inconsistencies.
- **Data Integration** is the integrating multiple databases, data cubes, or files, that is, data integration.
- **Data transformation** operations, such as normalization and aggregation, are additional data preprocessing procedures that would contribute toward the success of the mining process.
- **Data reduction** obtains a reduced representation of the data set that is much smaller in volume, yet produces the same (or almost the same) analytical results.

■ **Data Summarization**

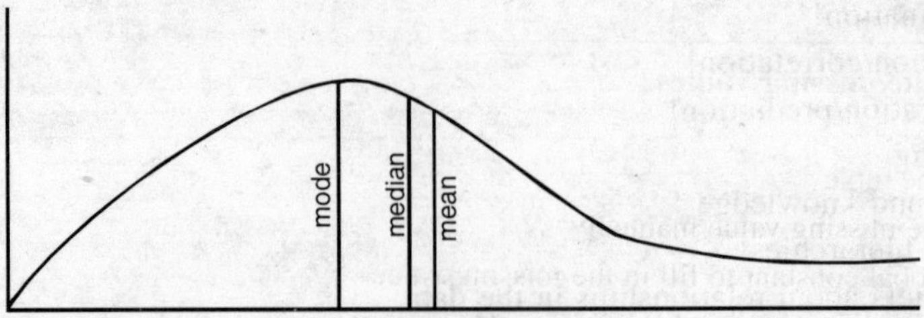

Fig. 7.1.

1. **Median** is the average of the middle two values.
2. A **holistic measure** is a measure that must be computed on the entire data set as a whole. It cannot be computed by partitioning the given data into subsets and merging the values obtained for the measure in each subset.

■ **Ways to measure the Dispersion of Data:**

The **range** of the set is the difference between the largest and smalles values.

The most commonly used percentiles other than the **median are quartiles.**

The distance between the first and third quartiles is a simple measure of spread that gives the range covered by the middle half of the data. This distance is called the interquartile range (IQR) and is defined as

IQR = Q3 - Q1.

Rule of thumb for identifying suspected outliers is to single out values falling at least 1.5 × IQR above the third quartile or below the first quartile.

Graphic Displays of Basic Descriptive Data Summaries

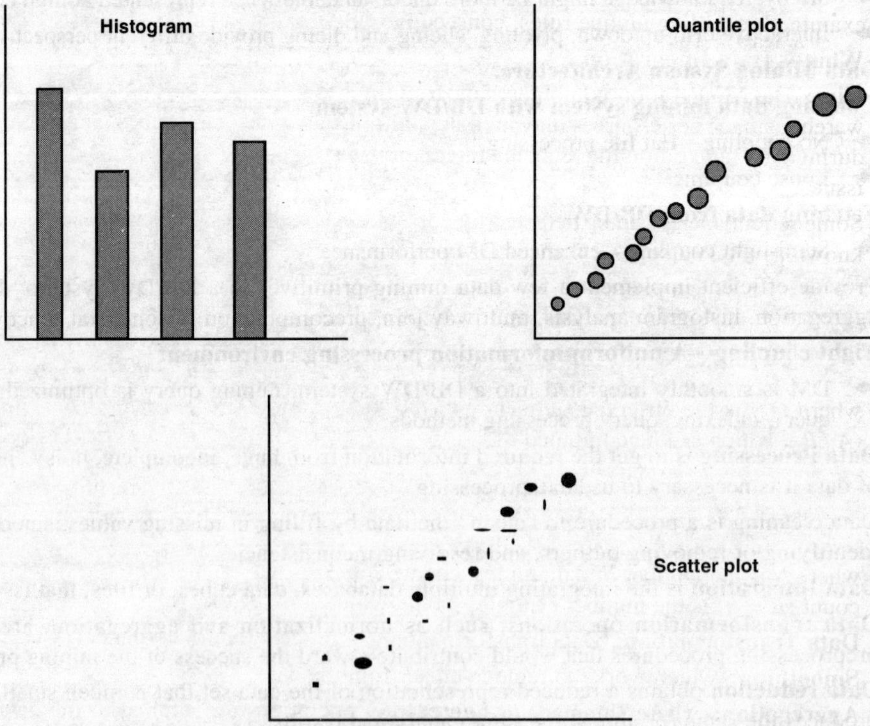

Fig. 7.2.

Review Questions

Q. 1. How to handle missing Values?

Hint:

- Ignore the tuple
- Fill in the missing value manually
- Use a global constant to fill in the missing value
- Use the attribute mean to fill in the missing value
- Use the attribute mean for all samples belonging to the same class as the given tuple
- Use the most probable value to fill in the missing value

Q. 2. How to handle Noisy Data?

Hint: Noise is a random error or variance in a measured variable. Given a numerical attribute such as, say, marks, we can **"smooth"** the data to remove the noise using Smoothing techniques like :

Binning: Binning methods smooth a sorted data value by consulting its "neighborhood," that is, the values around it. The sorted values are distributed into a number of "buckets," or bins. Because binning methods consult the neighborhood of values, they perform local smoothing

Regression: Data can be smoothed by fitting the data to a function, such as with regression. Linear regression involves finding the "best" line to fit two attributes (or variables), so that one attribute can be used to predict the other. Multiple linear regression is an extension of linear regression, where more than two attributes are involved and the data are fit to a multidimensional surface.

Clustering: Outliers may be detected by clustering, where similar values are organized into groups, or "clusters." Intuitively, values that fall outside of the set of clusters may be considered outliers.

The process of Data Cleaning: The first step in data cleaning as a process is discrepancy detection. knowledge or "data about data" is referred to as metadata is starting point. The data should also be examined regarding unique rules, consecutive rules, and null rules.

Q. 3. What is Data Integration ?

Hint: Data Integration combines data from multiple sources into a coherent data store, as in data warehousing. These sources may include multiple databases, data cubes, or flat files. Issues that arises during data integration like Schema integration and object matching Redundancy is another important issue.

Some redundancies can be detected by correlation analysis correlation coefficient. The ÷2 value (also known as the Pearson ÷ 2 statistic) is computed as:

$$\chi^2 = \sum_{i=1}^{c}\sum_{j=1}^{r}\frac{(o_{ij}-e_{ij})^2}{e_{ij}},$$

where $O_i j$ is the observed frequency of the joint event (A_i, B_j) and E_{ij} is the expected frequency of (A_i, B_j), which can be computed as

$$e_{ij} = \frac{count(A = a_i) \times count(B = b_j)}{N},$$

where N is the number of data tuples, count($A = ai$) is the number of tuples having value a for A, and count ($B = b$) is the number of tuples having value b for B.

Data Transformation: Data transformation can involve the following:

Smoothing: which works to remove noise from the data.

Aggregation: where summary or aggregation operations are applied to the data. For example, the daily sales data may be aggregated so as to compute weekly and annual total scores.

Generalization of the data: Where low-level or "primitive" (raw) data are replaced by higher-level concepts through the use of concept hierarchies. For example, categorical attributes, like street, can be generalized to higher-level concepts, like city or country.

Normalization: Where the attribute data are scaled so as to fall within a small specified range, such as –1.0 to 1.0, or 0.0 to 1.0.

Attribute construction: This is where new attributes are constructed and added from the given set of attributes to help the mining process.

Data Reduction techniques: These techniques can be applied to obtain a reduced representation of the data set that is much smaller in volume, yet closely maintains the integrity of the original data.

Strategies for data reduction include the following:

1. Data cube aggregation, where aggregation operations are applied to the data in the construction of a data cube.
2. Attribute subset selection, where irrelevant, weakly relevant, or redundant attributes or dimensions may be detected and removed.
3. Dimensionality reduction, where encoding mechanisms are used to reduce the data set size. Wavelet Transforms Principal Components Analysis
4. Numerosity reduction, where the data are replaced or estimated by alternative, smaller data representations such as parametric models (which need store only the model parameters instead of the actual data) or nonparametric methods such as clustering, sampling, and the use of histograms.
5. Discretization and concept hierarchy generation, where raw data values for attributes are replaced by ranges or higher conceptual levels. Data discretization is a form of numerosity reduction that is very useful for the automatic generation of concept hierarchies. Discretization and concept hierarchy generation are powerful tools for data mining, in that they allow the mining of data at multiple levels of abstraction.

Q. 4. Data Discretization and Concept Hierarchy Generation.

Hint: Data discretization techniques can be used to reduce the number of values for a given continuous attribute by dividing the range of the attribute into intervals. Interval labels can then be used to replace actual data values. Replacing numerous values of a continuous attribute by a small number of interval labels thereby reduces and simplifies the original data. This leads to a concise, easy-to-use, knowledge-level representation of mining results

Concept Hierarchy Generation for Categorical Data: Following are methods for generation of concept hierarchies for categorical data.

- Specification of a partial ordering of attributes explicitly at the schema level by users or experts.
- SpecificationSpecification of a set of attributes, but not of their partial ordering of a portion of a hierarchy by explicit data grouping.
- Specification of a set of attributes, but not of their partial ordering.

Example: Concept hierarchy generation using prespecified semantic connections. Suppose that a data mining administrator has pinned together the five attributes name, age, college, degree, grade because they are closely linked semantically regarding the notion of student. If a user were to specify only the attribute college for a hierarchy defining student, the system can automatically drag in all of the above five semantically related attributes to form a hierarchy.

❑❑❑

8

Chapter

Classification

INSIDE THIS CHAPTER

8.1. Introduction **8.2.** Classification and Issues **8.3.** Neural Networks **8.4.** Distance based Algorithms **8.5.** Case Based Reasoning (CBR) **8.6.** Statistical Based Algorithm (Bayesian Classification) **8.7.** Naive Bayes Classifier **8.8.** Bayesian Network **8.9.** Decision Tree Based Algorithm (ID3, C4.5, CART) **8.10.** Genetic Algorithm **8.11.** Rough Sets **8.12.** Fuzzy Sets Approach **8.13.** Support Victor Machine

8.1. INTRODUCTION

As described in previous chapter that the primary goal of data mining are : Predictive tasks and descriptive tasks. **Prediction or predictive tasks** involves using some variables or fields in the database to predict unknown or future values of other variables of interest and **Description or descriptive tasks** focuses on finding human-interpretable patterns describing the data.

The goals of prediction and description are achieved by using the following primary **data mining tasks** like **Classification, Regression, Clustering, Dependency Modeling, Summarization, Change and deviation detection.**

In this chapter the basic techniques for data classification such as NN, Genetic algorithm, Decision tree will be covered in more detail.

8.2. CLASSIFICATION AND ISSUES

Classification is a data mining function that assigns items in a collection to target categories or classes. The goal of classification is to accurately predict the target class for each case in the data. For example, a classification model could be used to identify loan applicants as low, medium, or high credit risks. Classifications look at the behaviour and attributes of already determined groups. The groups might include frequent flyers, high spenders, loyal customers, people responding to direct mail campaigns, or people with frequent back problems (people driving long distances every day). The data-mining tool can assign classifications to new data by examining existing data that has already been classified and by using the results to infer a set of rules. The set of rules is then applied to any new data to be classified. This technique often uses supervised induction, which employs a small training set of already classified records to determine additional classes.

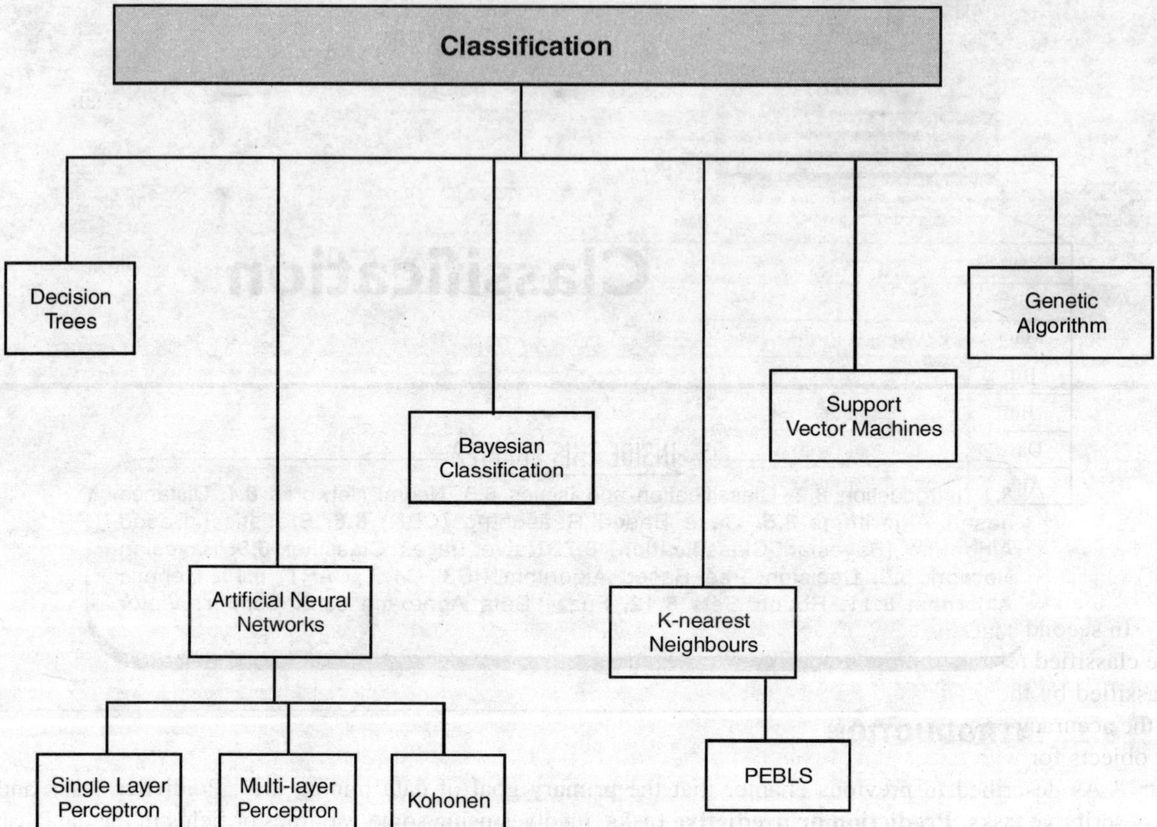

Fig. 8.1. Classification Techniques.

Classification is the task of mapping an input attribute set *x* into its class label *y*. Classification consists of two steps.

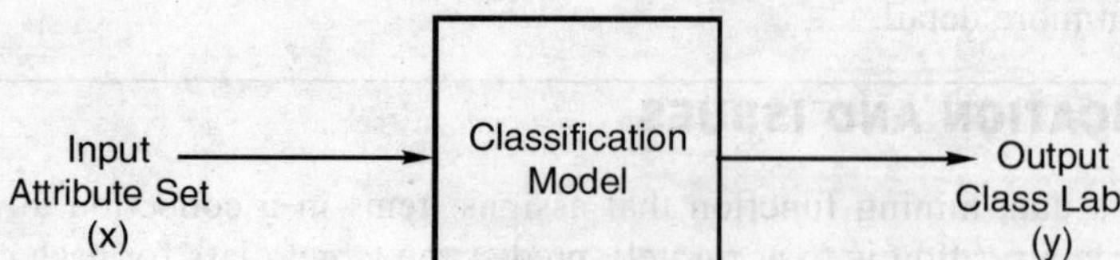

Fig. 8.2. Classification Model.

1. Model construction: Model construction deals with describing a set of predetermined classes.
2. Model usage: Model usage is for classifying future or unknown objects.

In first step the model is constructed that describe a predetermined set of data classes. Since the Information of classes is known in advance therefore this is an example of supervised learning. Each tuple/sample is assumed to belong to a predefined class, as determined by the class label attribute. The set of tuples that are used for model construction are called training set. The model is represented as classification rules, decision trees, or mathematical formulae.

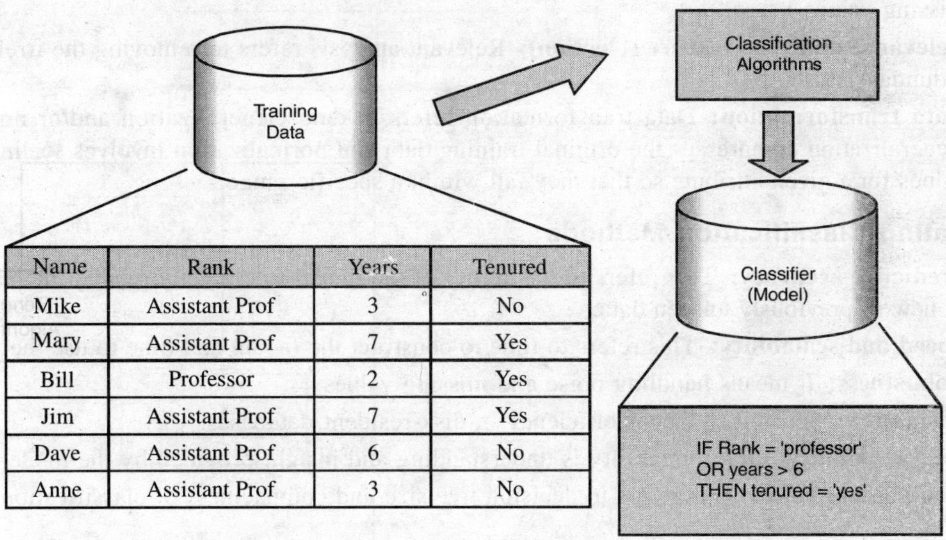

Name	Rank	Years	Tenured
Mike	Assistant Prof	3	No
Mary	Assistant Prof	7	Yes
Bill	Professor	2	Yes
Jim	Assistant Prof	7	Yes
Dave	Assistant Prof	6	No
Anne	Assistant Prof	3	No

Fig. 8.3. Classification Model for First Step.

In second step estimate the accuracy of the model. The known label of test sample is compared with the classified result from the model. Accuracy rate is the percentage of test set samples that are correctly classified by the model and the test set is independent of training set, otherwise over-fitting will occur. If the accuracy of the model is considered acceptable, the model can be used to classify future data tuples or objects for which class label is not known.

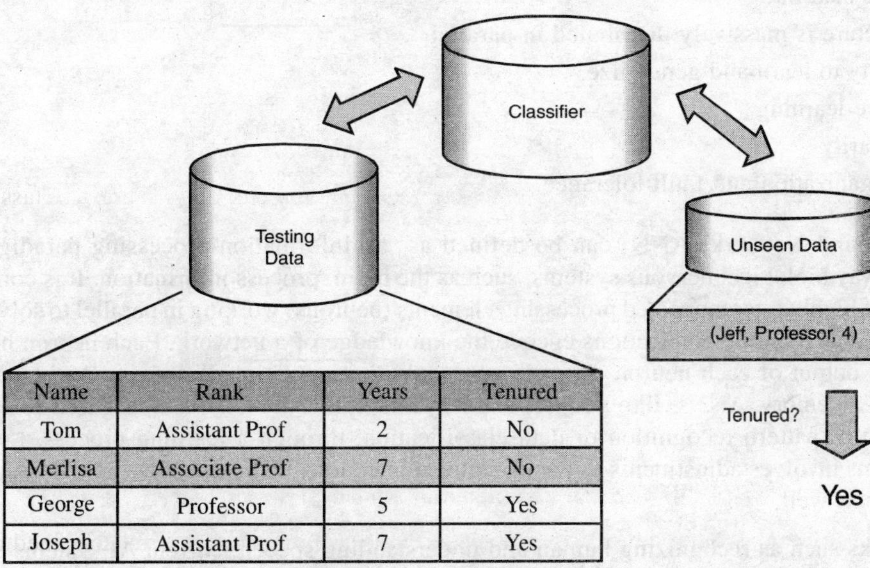

Name	Rank	Years	Tenured
Tom	Assistant Prof	2	No
Merlisa	Associate Prof	7	No
George	Professor	5	Yes
Joseph	Assistant Prof	7	Yes

Fig. 8.4. Classification Model for Second Step.

8.2.1. Issues

There are some issues regarding the data for classification described below.

1. Data Preparation

(*a*) **Data cleaning:** Data cleaning refers to preprocess data in order to reduce noise and handle missing values.

(*b*) **Relevance analysis (feature selection):** Relevant analysis refers to removing the irrelevant or redundant attributes.

(*c*) **Data transformation:** Data transformation refers to data Generalization and/or normalize. Generalization compresses the original training data and normalization involves scaling all the values for a given attribute so that they fall within a specific range.

2. Evaluating Classification Methods

(*a*) **Predictive accuracy:** This refers to the ability of the model to correctly predict the class label of new or previously unseen data.

(*b*) **Speed and scalability:** This refers to time to construct the model and time to use the model.

(*c*) **Robustness:** It means handling noise and missing values.

(*d*) **Scalability:** Scalability means efficiency in disk-resident databases.

(*e*) **Interpretability:** Interpretability is understanding and insight provided by the model.

(*f*) **Goodness of rules:** This refers to decision tree size and compactness of classification rules.

Since classification is an important task to map on input *x* to output *y*. For classification various techniques are available. Some of them are described below :

8.3. NEURAL NETWORKS [MDU Dec. 2011]

Neural Networks have been of greater interest while using DATA MINING algorithms. As we know that NN do have disadvantages like limitation in ease of use, ease of deployment etc. though they have some advantages also like

1. Its structure is massively distributed in parallel.
2. Its ability to learn and generalize.
3. Adaptive learning
4. No linearity
5. Self Organization and fault tolerance.

Artificial Neural Network (ANN) can be defined as an information-processing paradigm that is inspired by the way biological nervous systems, such as the brain, process information. It is composed of a large number of highly interconnected processing elements (neurons) working in parallel to solve specific problems. The weights on the connections encode the knowledge of a network. Each neuron has a local memory and the output of each neuron depends upon only the input signals arriving at the neuron and value in neuron's memory. ANNs, **like people**, learn by example. An ANN is configured for a specific application, such as pattern recognition or data classification, through a learning process. Learning in biological systems involves adjustments to the synaptic connections that exist between the neurons. This is true for ANNs as well.

In many tasks such as recognizing human and understanding speech, current AI systems cannot do better than humans. It is conjectured that the structure of the brain is somehow suited to these tasks and not suited to these tasks such as high speed arithmetic calculation. Rather than by being programmed an ANN can solve the problems simply by example mapping.

The intelligence of a neural network emerges from the collective behaviour of neurons, each of which performs only very limited operation. Even though each individual neuron works slowly, they can still quickly find the solution by working in parallel. This fact can explain why humans can recognize a visual scene faster than a digital computer, while an individual brain cell responds much more slowly than a digital cell in VLSI circuit.

8.3.1. Where to Use Artificial Neural Networks

One of the most common applications of NNs is in image processing. Some examples would be: identifying hand-written characters; matching a photograph of a person's face with a different photo in a database; performing data compression on an image with minimal loss of content. Other applications could be: voice recognition; RADAR signature analysis; stock market prediction. All of these problems involve large amounts of data, and complex relationships between the different parameters. Thus data mining came into picture for extracting useful information.

8.3.2. Neural Networks for Clustering

Neural networks of various kinds can be used for clustering and prototype creation. The Kohonen network described in this section is probably the most common network used for clustering and segmentation of the database. Typically the networks are used in a unsupervised learning mode to create the clusters. The clusters are created by forcing the system to compress the data by creating prototypes or by algorithms that steer the system toward creating clusters that compete against each other for the records that they contain, thus ensuring that the clusters overlap as little as possible.

8.3.3. Neural Networks for Feature Extraction

One of the important problems in all of data mining is that of determining which predictors are the most relevant and the most important in building models that are most accurate at prediction. These predictors may be used by themselves or they may be used in conjunction with other predictors to form "features". A simple example of a feature in problems that neural networks are working on is the feature of a vertical line in a computer image. The predictors, or raw input data are just the colored pixels that make up the picture. Recognizing that the predictors (pixels) can be organized in such a way as to create lines, and then using the line as the input predictor can prove to dramatically improve the accuracy of the model and decrease the time to create it.

In order to accomplish this the neural network tries to have the hidden nodes extract features from the input nodes that efficiently describe the record represented at the input layer. This forced "squeezing" of the data through the narrow hidden layer forces the neural network to extract only those predictors and combinations of predictors that are best at recreating the input record. The link weights used to create the inputs to the hidden nodes are effectively creating features that are combinations of the input nodes values.

8.3.4. How it Works

The NN approach like Decision trees require that a graphical structure be built to represent the model and then the structure be applied to the data. A neural network has a parallel-distributed architecture with a large no. of nodes and connections. There are 3 layers : input layer, hidden layer and output layer. Each connection points from one node to another and is associated with a weight. A simple view of the network structure and behaviours is shown in Fig. 8.5.

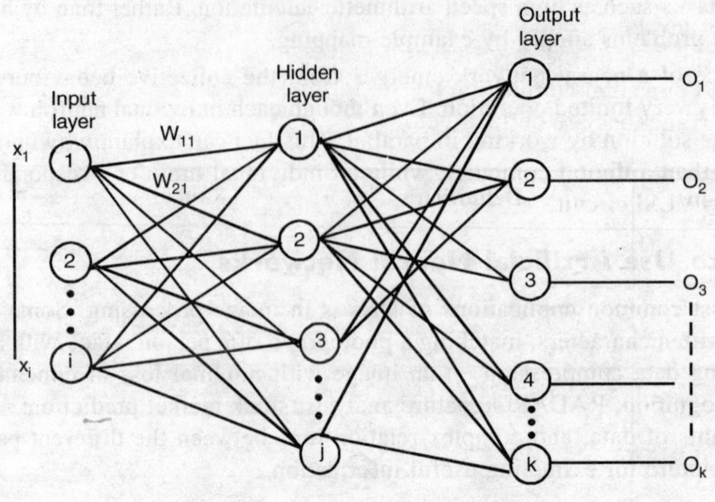

Fig. 8.5. The Neural Network Computation Model.

Neural Networks is one of the **Data Mining** techniques. To deal with any application:

The first step is to design a specific network architecture (that includes a specific number of "layers" each consisting of a certain number of "neurons"). The size and structure of the network needs to match the nature of the investigated phenomenon. Because the latter is obviously not known very well at this early stage, this task is not easy and often involves multiple "trials and errors."

The new network is then subjected to the process of "training." In that phase, neurons apply an iterative process to the number of inputs (variables) to adjust the weights of the network in order to optimally predict (in traditional terms, we could say find a "fit" to) the sample data on which the "training" is performed. After the phase of learning from an existing data set, the new network is ready and it can then be used to generate predictions.

The resulting *"network"* developed in the process of *"learning"* represents a pattern detected in the data.

Construction of NN for a particular application involves the following tasks:

- Determine the network properties: the network topologies, the type of connections, the order of connections and the weight range.
- Determine the node properties: the activation range, and the activation transfer function
- Determine the system dynamics: the weight initialization scheme, activation-calculating formula, and the learning rule.

1. Network Properties

The topology of a NN refers to its framework as well as its interconnection scheme. The framework is often specified by the number of layers and the number of nodes per layer. The types of layer include:

(*a*) **The input layer:** The nodes in it are called input units, which encodes the instance present to the network for processing, *e.g.,* each input unit may be designated by an attribute value possessed by the instance.

(*b*) **The hidden layer:** The nodes in it are called hidden units, which are not directly observable and hence hidden. They provide the nonlinearities for the network.

(*c*) **The output layer:** The nodes in it are called output nodes, which encodes possible concepts or values to be assigned to the instance under consideration. *e.g.,* each output unit presents a class of objects.

2. Node Properties

The activation levels of a node can be discrete (0 or 1) or continuous across the range ([0, 1]) or unrestricted. This depends on the activation function chosen. If it is hard limiting function then activation levels are 0 and 1. For a sigmoid function the activation levels are limited to a continuous range of [0, 1]. Various activation functions are defined below:

Function	Definition	Range	
Step/Threshold	$f(v) = \begin{cases} 1 & \text{if } v \geq 0 \\ 0 & \text{if } v < 0 \end{cases}$	[0, +1]	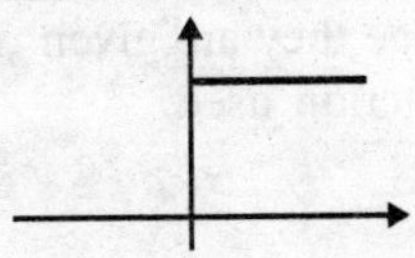
Signum	$f(v) = \begin{cases} 1 & \text{if } v > 0 \\ 0 & \text{if } v = 0 \\ -1 & \text{if } v < 0 \end{cases}$	[0, +1]	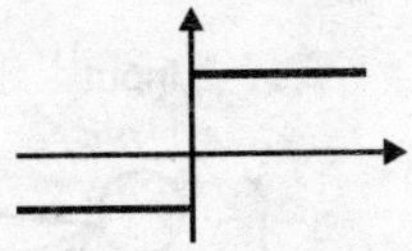
Linear Piecewise	$f(v) = \begin{cases} 1 & \text{if } v \geq 0 \\ v & \text{if } 0.5 > v > -0.5 \\ 0 & \text{if } v < -0.5 \end{cases}$	[0, +1]	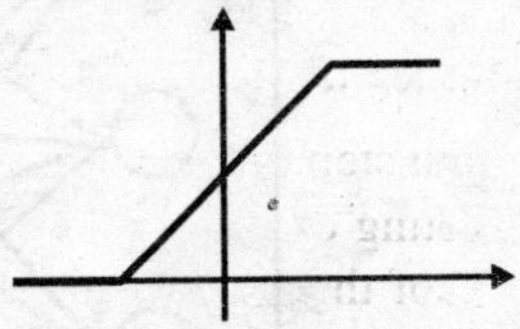
Identity	v	[–inf, +int]	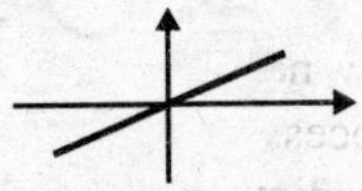
Sigmoidal Logistic	$f(v) = \dfrac{1}{1 + e^{-av}}$	[0, +1]	
Sigmoidal Hyperbolic	$f(v) = \tanh(v) = \dfrac{e^{v} - e^{-v}}{e^{v} + e^{-v}}$	[–1, +1]	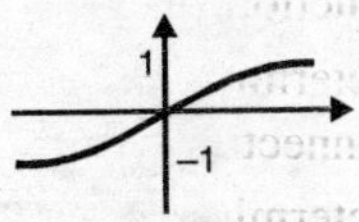
Gaussian	$f(v) = e^{-v}$	[–1, +1]	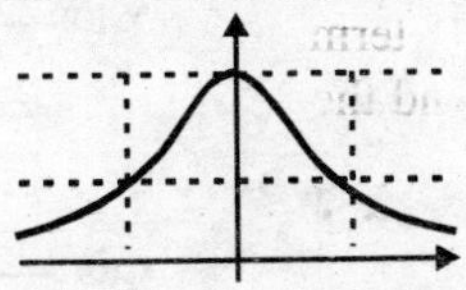

3. System Dynamics

- ➡ The weight initialization scheme is specific to the particular neural network model chosen. However, in many cases, initial weights are just randomized to small real numbers.
- ➡ The learning rule is one of the most important attributes to specify for a neural network. The learning rule determines how to adapt connection weights in order to optimize the network performance. It indicates how to calculate the weight adjustment during each training cycle. However, the rule is suspended after training is completed. Learning may be supervised, unsupervised or reinforced.

➡ When the neural network is used to solve the problem, the solution lies in the activation levels of the output units *e.g.*, suppose a NN is implemented for classifying fruits into lemons, oranges, apples. The network has three output units representing the three kinds respectively. Given an unknown fruit, we want to classify it. So we present the characteristics of the fruit to the network. The information is received by the input layer and propagates further. If the output unit corresponding to the class apple reaches the maximal activation, then the class assigned to the fruit is apple. From this example, the inference behaviour of a NN involves how to compute the activation levels across the network. The activation levels of the input units need not be calculated since they are given. Those of hidden and output units are calculated according to the activation function used.

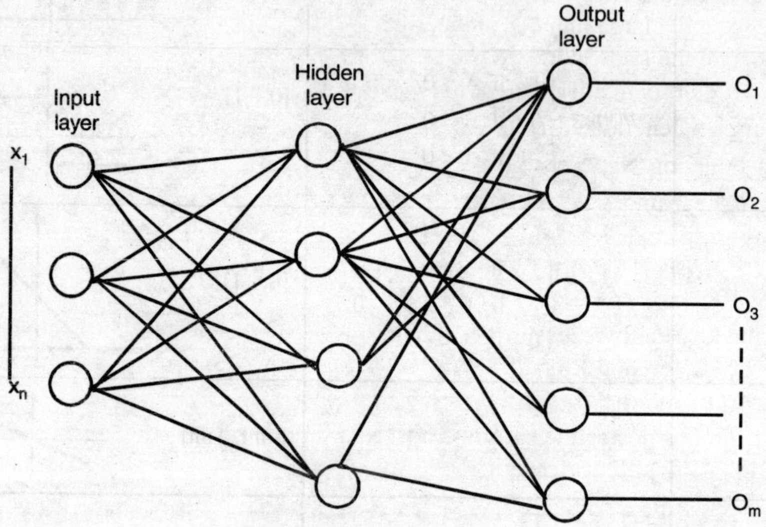

Fig. 8.6. Feedforward Network.

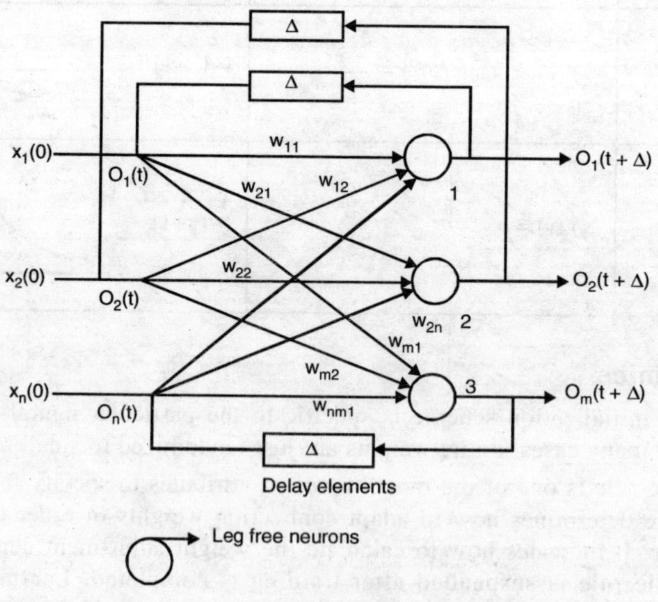

Fig. 8.7. Single Layer Discrete Time Feedback Networks.

8.3.5. Models of Artificial Neural Networks

ANN is defined as an interconnection of neurons such that the neuron outputs are connected, through the weights to all other neurons including themselves ; both lag-free and delay connections are allowed. In Feedforward Network, Data enters at the inputs and passes through the network, layer by layer, until it arrives at the outputs. During normal operation, that is when it acts as a classifier, there is no feedback between layers. This is why they are called *feedforward* neural networks. Alternatively a NN may be feedback where some links are back to earlier layers.

8.3.6. Neural Network Learning

The neural network has also been called the "connectionist". It contains a large number of neuron like processing elements and a large number of weighted connections between the elements. The weights on the connections encode the knowledge of a network. It uses a highly parallel, distributed control and can learn to adjust itself automatically. The various learning algorithms are described below:

1. Backpropagation (Generalized Delta Learning Rule)
2. Radial Basis Function Networks
3. Reinforcement Learning
4. ART Networks

1. Backpropagation

It was first described by Paul Werbos in 1974, but it wasn't until 1986, through the work of David E. Rumelhart, Geoffrey E. Hinton and Ronald J. Williams, that it gained recognition.

➡ Backpropagation network is **Multilayer feedforward network** with different transfer function in ANN and a more powerful learning rule. The learning rule is called **"Backpropagation"**, which is a kind of **gradient descent technique** with backward error propagation.

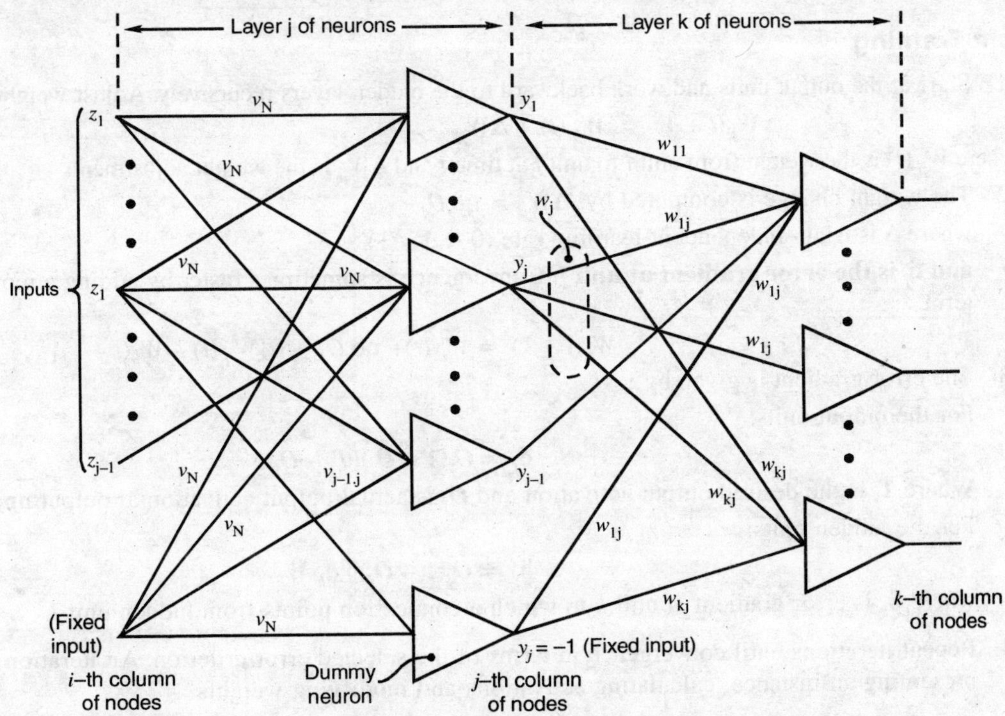

Fig. 8.8. Multilayer Network.

- It is a **Supervised Learning** method, and is an implementation of the **Delta rule**. It requires a teacher that knows, or can calculate, the desired output for any given input.

Conversely, here are some situations where a BP NN might be a good idea:

- A large amount of input/output data is available, but you're not sure how to relate it to the output.
- The problem appears to have overwhelming complexity, but there is clearly a solution.
- It is easy to create a number of examples of the correct behaviour.
- The solution to the problem may change over time, within the bounds of the given input and output parameters (*i.e.*, today 2 + 2 = 4, but in the future we may find that 2 + 2 = 3.8).
- Outputs can be "fuzzy", or non-numeric.

The Backpropagation Algorithm

Weight Initialization

Set all the weights and node thresholds to small random numbers. But node threshold is the negative of the weight from the bias unit.

Calculation of Activation

1. The activation level of an input unit is determined by the instances presented to the network.
2. The activation level O_j of hidden and output unit is determined by

$$O_j = F(\Sigma W_{ji} O_i - \theta_j)$$

Where W_{ji} is the weight from an input O_j, θ_j is the node threshold, and F is sigmoid function.

$$F(a) = \frac{1}{(1+e^{-a})} \qquad \text{where } a = \text{weight} * \text{input}$$

Weight Training

1. Start at the output units and work backward to the hidden layers recursively. Adjust weights by

$$\mathbf{W_{ji}(t+1) = W_{ji}(t) + \Delta W_{ji}}$$

where $W_{ji}(t)$ is the weight from unit i to unit j at time t and ΔW_{ji} is the weight adjustment.

2. The weight change is computed by $\mathbf{\Delta W_{ji} = \eta \delta_j O_i}$

   where Δ is trial –independent learning rate ($0 < \Delta < 1$)

   and δ_j is the error gradient at unit j. Convergence is sometimes faster by adding a momentum term:

$$W_{ji}(t+1) = W_{ji}(t) + \eta\delta_j O_i + \alpha[W_{ji}(t) - W_{ji}(t-1)]$$

3. The error gradient is given by :

   For the output units:

$$\mathbf{\delta_j = O_j(1 - O_j)(T_j - O_j)}$$

   Where T_j is the desired output activation and O_j is actual output activation at output unit j.

   For the hidden units:

$$\mathbf{\delta_j = O_j(1 - O_j)\Sigma\delta_k W_{kj}}$$

   where δ_k is error gradient at unit k to which a connection points from hidden unit j.

4. Repeat iterations until convergence in terms of the selected error criterion. An iteration include presenting an instance, calculating activations and modifying weights.

NOTE : The name "Backpropagation" comes from the fact that the error of hidden units are derived from propagating backward the errors associated with output units since the target values for the hidden units are not given. In this activation function used is sigmoid function, which compresses the output value into the range between 0 and 1.

Why use Sigmoid Function : The sigmoid function is advantageous because it can accommodate large signals without saturation while allowing the passing of small signals without excessive attenuation. Also, it is a smooth function so that gradients can be calculated which are required for a gradient descent search.

How to Solve the Numerical Based on Backpropagation

Consider three layers input layer, hidden layer and output layer.

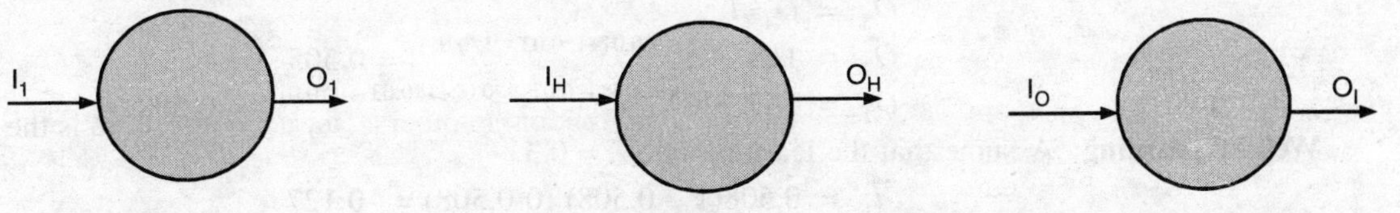

Fig. 8.9.

The input to the input layer is represented by I_I and the output of this layer $O_{I.}$

The input to the input layer is represented by I_H and the output of this layer $O_{H.}$

The input to the input layer is represented by I_O and the output of this layer $O_{O.}$

1. Write the input for input layer from the given question. The output of this layer is same as the input because there is no processing in the input layer.
2. Now calculate $I_{H.}$

   $I_H = [V]^t * O_I$. Where [V] is the weight between the input and the hidden layer.
3. Next is to calculate the output of the hidden layer O_H.

   $O_H = 1/[1 + \exp(-\lambda \text{net})$ where λ is a constant and net is the value calculated above. *i.e.*, net = weight*input for that layer.

   i.e., $O_H = 1/[1+\exp(-\lambda.\ I_H)$
4. Calculate I_O. $I_O = [W]^t * O_H$. Where [W] is the weight between the hidden layer and the output layer.
5. Calculate O_O. $O_O = 1/[1+\exp(-\lambda \text{net})$. Where λ is a constant and net is the value calculated above. *i.e.*, net = weight*input for that layer *i.e.*, $O_O = 1/[1 + \exp(-\lambda.\ I_O)$
6. Calculate the error by $E = [1/2]* (d - O)^2$.
7. Calculate error signal term for the output layer

$$\delta_0 = \frac{1}{2}\left[(d_k - o_k)(1 - o_k^2)\right]^*$$

8. Next Calculate error signal term for hidden layer $\delta_y = w_j^i \delta_0 f'_y$.

   Where $f'_y = O_H(1 - O_H)$.
9. Adjust the weights of the output layer

$$W^{\text{new}} = W^{\text{old}} + \eta\delta_0 . O_H^L$$

10. Adjust the weights of the hidden layer

$$V^{new} = V^{old} + \eta\delta_\gamma . I_H^L$$

Example :

1. To solve the EX-OR problem, we build a backpropagation network as shown in Fig. 8.8. The network will be trained on the following instances:

The weights are initialized randomly as follows:

$W_{13} = 0.02$, $W_{14} = 0.03$, $W_{12} = -0.02$, $W_{23} = 0.01$, $W_{24} = 0.02$, $W_{1b} = -0.01$, $W_{2b} = -0.01$

Calculation of Activation: Consider a training instance with the input vector = (1,1) and the desired output vector = 0.

$$O_3 = O_4 = 1$$
$$O_2 = 1/[1 + e^{-(1*0.01+1*0.02 - 1*0.01)}] = 0.505$$
$$O_1 = 1/[1 + e^{-(0.505* - 0.02+1*0.02+1*0.03 - 1*0.01)}] = 0.508$$

Weight Training: Assume that the learning rate $\lambda = 0.3$

$$d_1 = 0.508(1 - 0.508)(0\text{-}0.508) = -0.127$$
$$\text{Weight adjustment } \Delta W_{13} = \lambda d_1 O_3 = 0.3*(-0.127)*1 = -0.038$$
$$\lambda_2 = 0.505(1 - 0.505)(-0.127*\text{-}0.02) = -0.0006$$
$$\Delta W_{23} = 0.3*0.0006*1 = 0.0002$$

As the weight adjustments are decreasing therefore the rest of the weight adjustments are omitted.

2. Consider the simple network below :

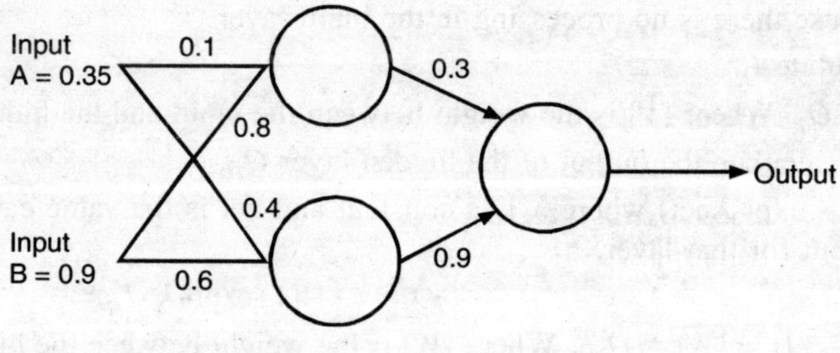

Fig. 8.10

Assume that the neurons have a Sigmoid activation function and

(*i*) Perform a forward pass on the network.

(*ii*) Perform a reverse pass (training) once (target = 0.5).

(*iii*) Perform a further forward pass and comment on the result.

Solution.

(*i*) Input to top neuron = (0.35*0.1) + (0.9*0.8) = 0.755.

Out = 1/1 + exp(–1*0.755) = 0.68.

Input to bottom neuron = (0.9*0.6) + (0.35*0.4) = 0.68. Out = 0.6637.

Input to final neuron = (0.3*0.68) + (0.9*0.6637) = 0.80133. Out = 0.69.

(*ii*) Output error $\delta = (t - o)(1 - o)o = (0.5 - 0.69)(1 - 0.69)0.69 = -0.0406$.

New weights for output layer

$$w1new = w1 + (\delta * \text{input}) = 0.3 + (-0.0406*0.68) = 0.272392.$$

$$w2new = w2+(\delta * input) = 0.9 + (-0.0406*0.6637) = 0.87305.$$

Errors for hidden layers:

$$\delta 1 = \delta * w1 = -0.0406 * 0.272392 * (1-o)o = -2.406*10-3$$

$$\delta 2 = \delta * w2 = -0.0406 * 0.87305 * (1-o)o = -7.916*10-3$$

New hidden layer weights:

$$w3^{new} = 0.1 + (-2.406 \text{ x } 10-3 \text{ x } 0.35) = 0.09916.$$

$$w4^{new} = 0.8 + (-2.406 \text{ x } 10-3 \text{ x } 0.9) = 0.7978.$$

$$w5^{new} = 0.4 + (-7.916 \text{ x } 10-3 \text{ x } 0.35) = 0.3972.$$

$$w6^{new} = 0.6 + (-7.916 \text{ x } 10-3 \text{ x } 0.9) = 0.5928.$$

(*iii*) Old error was –0.19. New error is –0.18205. Therefore error has reduced.

2. Radial Basis Function Networks (RBF)

A RBF network is a 2 layer network whose output unit forms a linear combination of the basis functions computed by hidden units. And each hidden unit implements a radial activated function. The output unit implements a weighted sum of hidden output units. The input to RBF network is nonlinear while the output is linear. Due to their nonlinear properties, RBF networks are able to model complex mappings. The most common basis function can be viewed as activation function in the hidden layer.

The most common basis function chosen is Gaussian function, in which case the activation level O_j of the hidden unit j is calculated as

$$O_j = \exp[-(X-W_j)^* (X-W_j)/2\sigma_j^2]$$

Difference between RBF and Backpropagation

The main difference is in their Basis Function. The radial basis function is used in RBF and it can cover only small regions, whereas the sigmoid function assumes large number of nonzero values over an infinitely large region of the input space. Depending upon the problem to be solved either can be used.

RBF NETWORK

1. Weight Initialization: weights in the hidden layer are determined by the clustering algorithm and weights in the output layer are initialized to small random values.

2. Calculation of activation: The activation level O_j of the hidden unit j is calculated as

$$O_j = \exp[-(X-W_j)^* (X-W_j)/2\sigma_j^2]$$

The activation level of an output unit is determined by

$$O_j = \Delta W_{ji} O_i$$

3. Weight Learning:

In the hidden layer, use k-means clustering algorithm

In the output layer:

(*a*) adjust the weights by $W_{ji}(t+1)= W_{ji}(t)+ \Delta W_{ji}(t)$

where $W_{ji}(t)$ is the weight from unit I to unit j at time j and $\Delta W_{ji}(t)$ is the weight adjustment.

(*b*) the weight change is computed by $\mathbf{\Delta W_{ji} = \Delta d_j O_i}$

where Δ is a trial independent learning rate and $\mathbf{d_j}$ is error at unit j :

$$\mathbf{d_j = T_j - O_j}$$

where $\mathbf{T_j}$ is desired output activation and $\mathbf{O_j}$ is actual output activation at output unit j .

(*c*) repeat the iterations until convergence.

8.4. DISTANCE BASED ALGORITHMS

k-nearest neighbour algorithm

The *k*-nearest neighbour algorithm (*k*-NN) is a method for classifying objects based on closest training examples in the feature space. *k*-NN is a type of instance-based learning, or lazy learning.

k is a positive integer, typically small, If $k = 1$, then the object is simply assigned to the class of its nearest neighbor.

There are 2 types of learning LAZY and EAGER

Lazy Learning	Eager Learning
Lazy learning (*e.g.*, instance-based learning): Simply stores training data (or only minor processing) and waits until it is given a test tuple. Lazy learning takes less time in training but more time in predicting.	**Eager learning :** Given a set of training tuples, constructs a classification model before receiving new (*e.g.*, test) data to classify
Accuracy: Lazy method effectively uses a richer hypothesis space since it uses many local linear functions to form an implicit global approximation to the target function.	**Accuracy :** Eager must commit to a single hypothesis that covers the entire instance space.
Examples: k-nearest neighbor approach, case based reasoning.	**Examples:** NN based, decision tree based etc.

The neighbours are taken from a set of objects for which the correct classification (or, in the case of regression, the value of the property) is known. This can be thought of as the training set for the algorithm, though no explicit training step is required. The *k*-nearest neighbor algorithm is sensitive to the local structure of the data.

Algorithm: The training examples are vectors in a multidimensional feature space, each with a class label.

1. Training phase: The training phase of the algorithm consists only of storing the feature vectors and class labels of the training samples.

2. Classification phase: In the classification phase, *k* is a user-defined constant, and an unlabelled vector (a query or test point) is classified by assigning the label which is most frequent among the *k* training samples nearest to that query point.

Usually Euclidean distance is used as the distance metric; however this is only applicable to continuous variables. In cases such as text classification, another metric such as the overlap metric (or Hamming distance) can be used. Often, the classification accuracy of "k"-NN can be improved significantly if the distance metric is learned with specialized algorithms such as Large Margin Nearest Neighbor or Neighbourhood components analysis.

A drawback to the basic "majority voting" classification is that the classes with the more frequent examples tend to dominate the prediction of the new vector, as they tend to come up in the *k* nearest neighbors when the neighbors are computed due to their large number. One way to overcome this problem is to weigh the classification taking into account the distance from the test point to each of its *k* nearest neighbors.

Nearest neighbour classifiers can also be used for prediction that is to return a real valued prediction for a given unknown sample. In this case the classifier returns the average value of the real valued labels associated with k nearest neighbors of the unknown sample.

8.5. CASE BASED REASONING (CBR)

CBR uses a database of problem solutions to solve new problems. CBR classifiers are also instance-based. CBR stores symbolic description (tuples or cases) not points in a Euclidean space like in K-NN algorithm.

Methodology

- First the instances are represented by rich symbolic descriptions (*e.g.*, function graphs)
- Then search for similar cases, multiple retrieved cases may be combined
- Tight coupling between case retrieval, knowledge-based reasoning, and problem solving

Challenges

- Find a good similarity metric
- Indexing based on syntactic similarity measure, and when failure, backtracking, and adapting to additional cases.

Applications

Customer-service (product-related diagnosis), legal ruling

8.6. STATISTICAL BASED ALGORITHM (BAYESIAN CLASSIFICATION)

8.6.1. Bayesian Classification

Bayesian classifiers are the statistical classifiers and can predict class membership probabilities such as the probability that a given sample belongs to a particular class. This classification is based on bayes theorem that is described below. Bayesian classifiers exhibit high accuracy and speed when applied to large databases.

8.6.2. Bayes Theorem

Bayes' theorem shows the relation between two **conditional probabilities** which are the reverse of each other. This theorem is named for **Thomas Bayes** and often called **Bayes' law** or **Bayes' rule**. Bayes' theorem expresses the conditional probability, or "posterior probability", of a hypothesis H (*i.e.*, its probability after evidence E is observed) in terms of the "prior probability" of H, the prior probability of E, and the conditional probability of E given H. It implies that evidence has a stronger confirming effect if it was more unlikely before being observed.

This states:

$$P(H_i|E) = \frac{P(E|H_i)\,P(H_i)}{\sum_{k=1}^{n} P(E|H_k)\,P(H_k)}$$

This can be read as that given some evidence E then probability that hypothesis H_i is true is equal to the ratio of the probability that E will be true. Given H_i times the *a priori* evidence on the probability of H_i and the sum of the probability of E over the set of all hypotheses times the probability of these hypotheses.

- The set of all hypotheses **must** be mutually exclusive and exhaustive.
- Thus to find if we examine medical evidence to diagnose an illness. We must know all the prior probabilities of find symptom and also the probability of having an illness based on certain symptoms being observed.

Simple Statement of Theorem

Thomas Bayes addressed both the case of discrete probability distributions of data and the more complicated case of continuous probability distributions. In the discrete case, Bayes' theorem relates the conditional and marginal probabilities of events *A* and *B*, provided that the probability of *B* does not equal zero:

$$P(A \mid B) = \frac{P(B \mid A)\, P(A)}{P(B)}$$

In Bayes' theorem, each probability has a conventional name:

- *P*(*A*) is the *prior* probability (or "*unconditional*" or "*marginal*" probability) of *A*. It is "prior" in the sense that it does not take into account any information about *B*; however, the event *B* need not occur after event *A*.
- *P*(*A*|*B*) is the conditional probability of *A*, given *B*. It is also called the posterior probability because it is derived from or depends upon the specified value of *B*.
- *P*(*B*|*A*) is the conditional probability of *B* given *A*. It is also called the likelihood.
- *P*(*B*) is the prior or marginal probability of *B*, and acts as a normalizing constant.

Bayes' theorem in this form gives a mathematical representation of how the conditional probability of event *A* given *B* is related to the converse conditional probability of *B* given *A*.

8.7. NAIVE BAYES CLASSIFIER

The Naïve Bayes classifier works on a simple, but comparatively intuitive concept. Also, in some cases it is also seen that Naïve Bayes outperforms many other comparatively complex algorithms. It makes use of the variables contained in the data sample, by observing them individually, independent of each other.

The Naïve Bayes classifier is based on the Bayes rule of conditional probability. It makes use of all the attributes contained in the data, and analyses them individually as though they are equally important and independent of each other. For example, consider that the training data consists of various animals (say elephants, monkeys and giraffes), and our classifier has to classify any new instance that it encounters. We know that elephants have attributes like they have a trunk, huge tusks, a short tail, are extremely big, etc. Monkeys are short in size, jump around a lot, and can climb trees; whereas giraffes are tall, have a long neck and short ears.

The Naïve Bayes classifier will consider each of these attributes separately when classifying a new instance. So, when checking to see if the new instance is an elephant, the Naïve Bayes classifier will not check whether it has a trunk and has huge tusks and is large. Rather, it will separately check whether the new instance has a trunk, whether it has tusks, whether it is large, etc. It works under the assumption that one attribute works independently of the other attributes contained by the sample.

8.7.1. The Naive Bayes Probabilistic Model

Abstractly, the probability model for a classifier is a conditional model

$$p(C \mid F_1, \ldots, F_n)$$

over a dependent class variable C with a small number of outcomes or *classes*, conditional on several feature variables F_1 through F_n. The problem is that if the number of features n is large or when a feature can take on a large number of values, then basing such a model on probability tables is infeasible. We therefore reformulate the model to make it more tractable.

Using **Bayes' theorem**, we write

$$p(C \mid F_1, ..., F_n) = \frac{p(C)\, p(F_1, ..., F_n / C)}{p(F_1, ..., F_n)}$$

In plain English the above equation can be written as

$$\text{Posterior} = \frac{\text{Prior} \times \text{Likelihood}}{\text{Evidence}}$$

In practice we are only interested in the numerator of that fraction, since the denominator does not depend on C and the values of the features F_i are given, so that the denominator is effectively constant. The numerator is equivalent to the **joint probability** model $p(C, F_1, ..., F_n)$ which can be rewritten as follows, using the **chain rule** for repeated applications of the definition of **conditional probability**:

$$p(C, F_1, ..., F_n)$$

$$\propto p(C)\, p(F_1, ..., F_n \mid C)$$

$$\propto p(C)\, p(F_1 \mid C)\, p(F_2, ..., F_n \mid C, F_1)$$

$$\propto p(C)\, p(F_1 \mid C)\, p(F_2 \mid C, F_1)\, p(F_3, ..., F_n \mid C, F_1, F_2)$$

$$\propto p(C)\, p(F_1 \mid C)\, p(F_2 \mid C, F_1)\, p(F_3 \mid C, F_1, F_2)\, p(F_4, ..., F_n \mid C, F_1, F_2, F_3)$$

$$\propto p(C)\, p(F_1 \mid C)\, p(F_2 \mid C, F_1)\, p(F_3 \mid C, F_1, F_2) ... p(F_n \mid C, F_1, F_2, F_3, ..., F_{n-1}).$$

Now the "naive" **conditional independence** assumptions come into play: assume that each feature F_i is conditionally **independent** of every other feature F_j for $j \neq i$. This means that

$$p(F_i \mid C, F_j) = p(F_i \mid C)$$

for $i \neq j$, and so the joint model can be expressed as

$$p(C, F_1, ..., F_n) \propto p(C)\, p(F_1 \mid C)\, p(F_2 \mid C)\, p(F_3 \mid C) ...$$

$$\propto p(C) \prod_{i=1}^{n} p(F_i \mid C).$$

This means that under the above independence assumptions, the conditional distribution over the class variable C can be expressed like this:

$$p(C \mid F_1, ..., F_n) = \frac{1}{Z} p(C) \prod_{i=1}^{n} p(F_i \mid C)$$

where Z (the evidence) is a scaling factor dependent only on $F_1, ..., F_n$, *i.e.*, a constant if the values of the feature variables are known.

Models of this form are much more manageable, since they factor into a so-called *class prior* $p(C)$ and independent probability distributions $p(F_i \mid C)$. If there are k classes and if a model for each $p(F_i \mid C = c)$ can be expressed in terms of r parameters, then the corresponding naive Bayes model has $(k - 1) + n\, r\, k$ parameters. In practice, often $k = 2$ (binary classification) and $r = 1$ (**Bernoulli**

variables as features) are common, and so the total number of parameters of the naive Bayes model is $2n + 1$, where n is the number of binary features used for classification and prediction.

If one is dealing with continuous data, a typical assumption is that the continuous values associated with each class are distributed according to a Gaussian distribution.

For example, suppose the training data contain a continuous attribute, x. We first segment the data by the class, and then compute the mean and **variance** of x in each class. Let μ_c be the mean of the values in x associated with class c, and let σ_c^2 be the variance of the values in x associated with class c. Then, the probability of some value given a class, $P(x = v \mid c)$, can be computed by plugging v into the equation for a Normal distribution parameterized by μ_c and σ_c^2. That is,

$$P(x = v \mid c) = \frac{1}{\sqrt{2\pi\sigma_c^2}} e^{\frac{(v-\mu_c)^2}{2\sigma_c^2}}$$

Another common technique for handling continuous values is to use binning to **discretize** the values. In general, the distribution method is a better choice if there is a small amount of training data, or if the precise distribution of the data is known. The discretization method tends to do better if there is a large amount of training data because it will learn to fit the distribution of the data. Since naive Bayes is typically used when a large amount of data is available (as more computationally expensive models can generally achieve better accuracy), the discretization method is generally preferred over the distribution method.

8.7.2. Bayes' Theorem Derived via Conditional Probabilities

To derive Bayes' theorem, start from the definition of **conditional probability**. The probability of the event A given the event B is

$$P(A \mid B) = \frac{P(A \cap B)}{P(B)}$$

Equivalently, the probability of the event B given the event A is

$$P(B \mid A) = \frac{P(A \cap B)}{P(A)}$$

Rearranging and combining these two equations, we find

$$P(B \mid A)\, P(B) = P(A \cap B) = P(B \mid A)\, P(A)$$

This **lemma** is sometimes called the product rule for probabilities. Discarding the middle term and dividing both sides by $P(B)$, provided that neither $P(B)$ nor $P(A)$ is 0, we obtain Bayes' theorem:

$$P(A \mid B) = \frac{P(B \mid A)\, P(A)}{P(B)}$$

Of course, this lemma is symmetric in A and B, since A and B are arbitrarily-chosen symbols, and dividing by $P(A)$, provided that it is non-zero, gives a statement of Bayes' theorem in which the two symbols have changed places.

Alternative Forms

Bayes' theorem is often completed by noting that, according to the **Law of total probability**,

$$P(B) = P(A \cap B) + P(A^c \cap B) = P(B|A)\,P(A) + P(B|A^c)\,P(A^c)$$

where A^c is the **complementary** event of A (often called "not A").

This results in the analogous form:

$$P(A \mid B) = \frac{P(B|A)\,P(A)}{P(B|A)\,P(A) + P(B|A^c)\,P(A^c)}$$

More generally, the law states that given a **partition**, *i.e.*, $\{A_i\}$, of the event space,

$$P(B) = \sum_i P(B \cap A_i) = \sum_i P(B|A_i)\,P(A_i)$$

Thus, for any A_i in the partition, Bayes' theorem states that

$$P(A_i \mid B) = \frac{P(B|A_i)\,P(A_i)}{P(B)} = \frac{P(B|A_i)\,P(A_i)}{\sum_j P(B|A_j)\,P(A_j)}$$

In terms of odds and likelihood ratio.

Bayes' theorem can also be written neatly in terms of a likelihood ratio Λ and **odds** O as

$$O(A \mid B) = O(A)\,.\,\Lambda(A \mid B)$$

where $O(A \mid B)$ are the (posterior) *odds* of A given B,

$$O(A \mid B) = \frac{P(A|B)}{P(A^c \mid B)}$$

$O(A)$ are the (prior) odds of A by itself

$$O(A) = \frac{P(A)}{P(A^c)}$$

and $\Lambda(A \mid B)$ is the *likelihood ratio.*

$$L(A \mid B) = \frac{P(B|A)}{P(B|A^c)}$$

Extensions

Theorems analogous to Bayes' theorem cover more than two events. For example:

$$P(A|B \cap C) = \frac{P(A)\,P(B|A)\,P(C|A \cap B)}{P(B)\,P(C|B)}.$$

This can be derived in a few steps from Bayes' theorem and the definition of conditional probability:

$$P(A|B \cap C) = \frac{P(A \cap B \cap C)}{P(B \cap C)}$$

$$= \frac{P(C|A \cap B)\,P(A \cap B)}{P(B)\,P(C|B)} = \frac{P(A)\,P(B|A)\,P(C|A \cap B)}{P(B)\,P(C|B)}$$

Similarly,

$$P(A|B \cap C) = \frac{P(B|A \cap C)\,P(A|C)}{P(B|C)}$$

can be regarded as a conditional Bayes' Theorem and can be derived as follows:

$$P(A \mid B \cap C) = \frac{P(A \cap B \cap C)}{P(B \cap C)}$$

$$= \frac{P(B \mid A \cap C)\, P(A \mid C)\, P(C)}{P(C)\, P(B \mid C)}$$

$$= \frac{P(B \mid A \cap C)\, P(A \mid C)}{P(B \mid C)}$$

Example: *The Monty Hall problem*

Three doors were presented — red, green, and blue — from which to choose only one, one of which has a prize hidden behind it. Suppose we choose the red door. The host of the contest, who *knows the location of the prize* and *will not open that door*, opens the blue door and reveals that there is no prize behind it. He then asks if we wish to change from our initial choice of red. Will changing to green now improve our chances of winning the prize?

Solution. Consider that two doors are left unopened, so there is 50:50 chance with either one, therefore there is no point for or against changing doors. However, this is not true.

Suppose the situations where the prize is behind one of the doors A_r, A_g, and A_b, for the red, green, and blue doors, respectively. It is commonly assumed the prize is placed randomly, hence

$$P(A_r) = P(A_g) = P(A_b) = \frac{1}{3}$$

Suppose "the host opens the blue door" proposition B. It is assumed the host choose at random when he has a choice, hence B must have probability 1/2.

When the prize is behind the red door, the host is free to open the green or the blue door. If the host opens them uniformly at random:

$$P(B \mid A_r) = \frac{1}{2}.$$

- When the prize is behind the green door, the host must open the blue door. Thus: $P(B \mid A_g) = 1$.
- When the prize is behind the blue door, the host must open the green door. Thus, $P(B \mid A_b) = 0$.

Thus, under the condition the red door is chosen, we get:

$$P(A_r \mid B) = \frac{P(B \mid A_r)\, P(A_r)}{P(B)} = \frac{\frac{1}{2} \cdot \frac{1}{3}}{\frac{1}{2}} = \frac{1}{3}$$

$$P(A_g \mid B) = \frac{P(B \mid A_g)\, P(A_g)}{P(B)} = \frac{1 \cdot \frac{1}{3}}{\frac{1}{2}} = \frac{2}{3}$$

$$P(A_b \mid B) = \frac{P(B \mid A_b)\, P(A_b)}{P(B)} = \frac{0 \cdot \frac{1}{3}}{\frac{1}{2}} = 0$$

So, we should change from the red to the green door for the higher probability of winning.

All of this assumes that after a door is chosen, the second door is opened at random, with equal probability, whenever two are available: the rooms behind both the green and blue doors are empty because behind the red door is the prize.

If the blue door is opened in that situation, it makes no difference what we do. If the blue door is never opened when the rooms behind both the green and blue doors are empty, then with *certainty* ,we will get the prize by changing from red to green. Under these two alternative assumptions, "always" and "never", the unconditional probability that the host opens the blue door happens to be 2/3 and 1/3, rather than 1/2 as above.

8.7.3. Certainty Factor

A certainty factor is based on measures of belief B and disbelief D of an hypothesis H_i given evidence E as follows:

$$B(H_i \mid E) = \begin{cases} 1 & \text{if } P(H_i) = 1 \\ \dfrac{\max[P(H_i \mid E), P(H_i)] - P(H_i)}{(1 - P(H_i))\, P(H_i \mid E)} & \text{otherwise} \end{cases}$$

$$D(H_i \mid E) = \begin{cases} 1 & \text{if } P(H_i) = 0 \\ \dfrac{P(H_i) - \min[P(H_i \mid E), P(H_i)]}{P(H_i)\, P(H_i \mid E)} & \text{otherwise} \end{cases}$$

where $P(H_i)$ is the standard probability.

The certainty factor C of some hypothesis H_i given evidence E is defined as:

$$C(H_i \mid E) = B(H_i \mid E) - D(H_i \mid E)$$

8.8. BAYESIAN NETWORK

A Bayesian network is a **belief network** or **directed acyclic graphical model** is a probabilistic graphical model that represents:

- A set of random variables and conditional dependencies via direct acyclic graph(DAG).
- A set of directed links connects pairs of nodes
- Each nodes represent random variables in the Bayesian sense: they may be observable quantities, latent variables, unknown parameters or hypotheses. The graph has no directed cycles (DAG).
- Edges represent conditional dependencies; nodes which are not connected represent variables which are conditionally independent of each other. Each node is associated with a probability function that takes as input a particular set of values for the node's parent variables and gives the probability of the variable represented by the node.
- It is usually much easier for an expert to decide conditional dependence relationships than specifying probabilities
- Sometimes, experts can have very different opinions
- Once the network is specified, we need only specify conditional probabilities for the nodes that participate in direct dependencies, and use those to compute any other probabilities.

Consider the following example:

The probability, $P(\delta_1)$

- ➡ represents that the car won't start.
- ➡ If the car won't start then it is likely that
  - ■ The battery is flat or
  - ■ The starting motor is broken.

In order to decide whether to fix the car myself or send it to the garage then following decisions can be made:

- ➡ If the headlights do not work then the battery is likely to be flat so *i* fix it myself.
- ➡ If the starting motor is defective then send car to garage.
- ➡ If battery and starting motor both gone send car to garage.

The network to represent this is as follows:

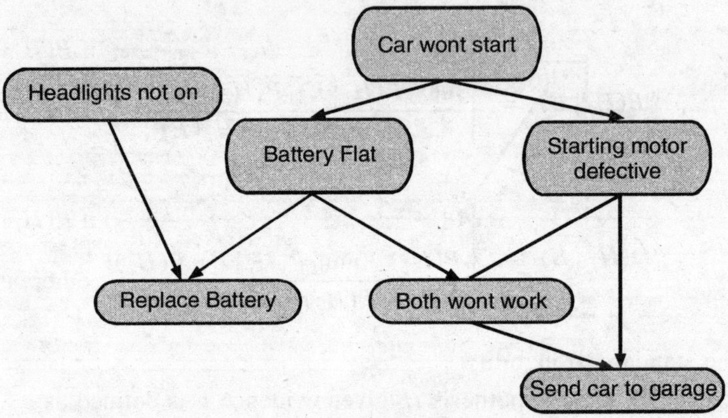

Fig. 8.11. A Simple Bayesian Network.

Training Bayesian Networks: Several Scenarios

Scenario 1: Given both the network structure and all variables observable: it consists of *computing only the CPT (conditional probability rate) entries.*

Scenario 2: Network structure known, some variables hidden: use the *gradient descent* (greedy hill-climbing) method, *i.e.,* search for a solution along the steepest descent of a criterion function

Weights are initialized to random probability values

At each iteration, it moves towards what appears to be the best solution at the moment, without backtracking.

Weights are updated at each iteration and converge to local optimum

Scenario 3: Network structure unknown, all variables observable: search through the model space to *reconstruct network topology*

Scenario 4: Unknown structure, all hidden variables: No good algorithms known for this purpose.

8.9. DECISION TREE BASED ALGORITHM (ID3, C4.5, CART)

What is a Decision Tree (DT)?

A decision tree is a predictive model that, as its name implies, can be viewed as a

- ➡ Tree where the root and each internal node is labelled with a question.

➡ The arcs represent each possible answer to the associated question.

➡ Each leaf node represents a prediction of a solution to the problem.

Decision tree learning is a method commonly used in data mining. The goal is to create a model that predicts the value of a target variable based on several input variables and the result is classification for that tuple. An example is shown here.

Each **interior node** corresponds to one of the input variables; there are edges to children for each of the possible values of that input variable. Each leaf represents a value of the target variable given the values of the input variables represented by the path from the root to the leaf.

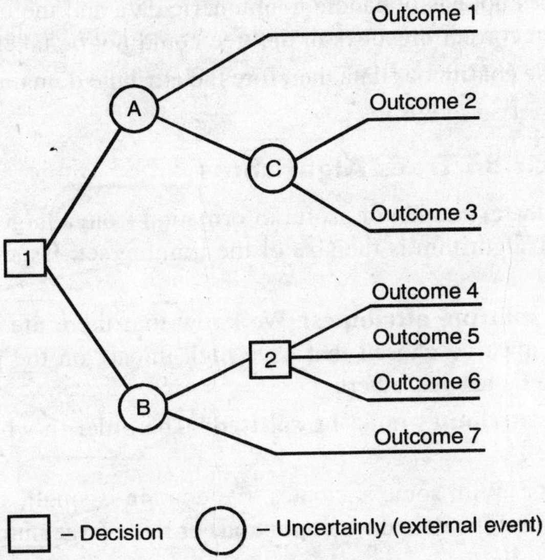

Fig. 8.12.

Squares represent decisions you can make. The lines that come out of each square on its right show all the available distinct options that can be selected at that decision analysis point.

Circles show various circumstances that have uncertain outcomes (For example, some types of events that may affect you on a given path). The lines that come out of each circle denote possible outcomes of that uncontrollable circumstance. Write down above each such line in the decision tree your best guesses for probabilities (for example, "80%" or "0.8") of those different outcomes.

Each path that can be followed along the decision tree, from left to right, leads to some specific outcome.

Decision trees used in **data mining** are of two main types:

➡ **Classification tree** analysis is when the predicted outcome is the class to which the data belongs.

➡ **Regression tree** analysis is when the predicted outcome can be considered a real number (*e.g.*, the price of a house, or a patient's length of stay in a hospital).

8.9.1. Advantages and Disadvantages of Decision Trees

There are many advantages and disadvantages to the use of decision trees for classification. These are:

Advantages:

➡ DTs are Easy to understand and easy to use.

➡ Rules can be generated that are easy to interpret and understand.

- Because tree size is independent of database size therefore they scale well for large databases.
- Since each tuple in the database must be filtered through the trees therefore this takes time proportional to the height of the tree .
- Trees can be constructed for data with many attributes.

Disadvantages:

- DTs may suffer from overfitting, And this problem can be overcome by tree pruning
- DT classifies by rectangular partitioning, but all the classification problems are not of this type.
- Decision tree does not easily handle nonnumeric data and the missing data. Because if there is missing data then correct branches in the tree could not be taken.
- DTs do not handle continuous data therefore the attribute domain must be divided into categories to be handled.

8.9.2. Issues in Decision Trees Algorithms

Since data mining is the extraction of useful information from a large database, but the major factors in the performance of DT algorithm is the size of the training set. Issues faced by most DT algorithms are:

1. **How to choose splitting attributes:** We know that there are many attributes in a database, which attributes must be chosen that have high impact on the performance is an issue. Some attributes may be better that others.
2. **In Which order attributes must be splitted:** The order in which the attributes are chosen is also important.
3. **Number of Splits:** With some attributes, the domain is small, so the number of splits is based on the domain. But if the domain is continuous or has a large number of splits to use is not easily determined.
4. **Stopping Criteria:** If the training data are perfectly classified then the termination of tree definitely occurs. But in some situations to prevent the creation of large trees the stopping criteria is desirable. If the training data is too small , then there may be a problem and the generated tree might not be specific enough to work properly with the more general data. And if the training data is too large it may lead to overfitting.
5. **Tree pruning:** When a DT is built , some modifications to the tree might be needed to improve the performance of the tree during the classification phase. Tree pruning methods address the overfitting problem. Tree pruning methods remove the redundant comparisons or remove the subtrees to achieve better performance. Two types of tree pruning are prepruning and post pruning. Techniques

Pruning can occur in a top down or bottom up fashion. A top down pruning will traverse nodes and trim subtrees starting at the root, while a bottom up pruning will start at the leaf nodes. Below are several popular pruning algorithms. Reduced error pruning

One of the simplest forms of pruning is reduced error pruning. Starting at the leaves, each node is replaced with its most popular class. If the prediction accuracy is not affected then the change is kept. While somewhat naive, reduced error pruning has the advantage of simplicity and speed.

Cost Complexity Pruning

Cost complexity pruning generates a series of trees $T_0 \ldots T_m$

where T_0 is the initial tree and T_m is the root alone. At step i the tree is created by removing a subtree from tree $i - 1$ and replacing it with a leaf node with value chosen as in the tree building algorithm. The subtree that is removed is chosen as follows. Define the error rate of tree *T* over data set *S* as err(*T*, *S*). The subtree that minimizes

$$\frac{err(prune(T,t),S)-err(T,S)}{|leaves(T)|-|leaves(prune(T,t))|}$$

is chosen for removal. The function prune(*T*, *t*) defines the tree gotten by pruning the subtrees *t* from the tree *T*. Once the series of trees has been created, the best tree is chosen by generalized accuracy as measured by a training set or cross-validation.

8.9.3. Decision Tree Based Algorithms

Various Decision tree based algorithms are:

8.9.3.1. ID3: ID3 stands for Iterative Dichotomiser 3

It is an Algorithm used to generate a decision tree. ID3 is a ancestor to the C4.5 Algorithm. The ID3 algorithm was invented by Ross Quinlan. Quinlan was a computer science researcher in data mining, and decision theory.

ID3 technique used to generate DT is based on Information theory and Attempts to create the smallest possible decision tree. The basic idea of this algorithm is to ask the questions whose answer provides the most information. The process is explained below:

1. Take all unused attributes and calculates their entropies. Entropy is used to measure the amount of uncertainty in a set of data. If the data is belong to only one class there is no entropy. A complete homogeneous sample has an entropy of 0 and An equally divided sample as an entropy of 1.
   Entropy = –p+log2 (p+) -p-log2 (p-) for a sample of negative and positive elements and
   $$\text{Entropy(S)} = \sum_{i=1}^{c} p_i \log_2 p_i$$
2. Chooses attribute that has the lowest entropy is minimum or when information gain is maximum. Information Gain is based on the decrease in entropy after a dataset is split on an attribute. Information gain is also said looking for which attribute creates the most homogeneous branches.
3. Makes a node containing that attribute.

Algorithm for Decision Tree

- Create a root node for the tree
- If all examples are positive, Return the single-node tree Root, with label = +.
- If all examples are negative, Return the single-node tree Root, with label = –.
- If number of predicting attributes is empty, then Return the single node tree Root, with label = most common value of the target attribute in the examples.
- Else
  - A = The Attribute that best classifies examples.
  - Decision Tree attribute for Root = A.
  - For each possible value, vi, of A.
  - Add a new tree branch below Root, corresponding to the test A = vi.
  - Let Examples(vi), be the subset of examples that have the value vi for A.
  - If Examples(vi) is empty.
  - Then below this new branch add a leaf node with label = most common target value in the examples.

- Else below this new branch add the subtree ID3 (Examples(vi), Target_Attribute, Attributes – {A})

➡ End

➡ Return Root

Example 8.1. Calculate the entropy. Given: Set S contains14 examples, 9 Positive values and 5 Negative values.

Solution.

$$\text{Entropy}(S) = -(9/14)\ \text{Log}_2\ (9/14) - (5/14)\ \text{Log}_2\ (5/14) = 0.940$$

Example 8.2. Information Gain 14 examples, 9 positive 5 negative

- **The attribute is Wind.**
- **Values of wind are Weak and Strong.**

Solution. 8 occurrences of weak winds

6 occurrences of strong winds

For the weak winds, 6 are positive and 2 are negative

For the strong winds, 3 are positive and 3 are negative

➡ Gain(S,Wind) =

➡ Entropy(S) - (8/14)*Entropy (Weak) -(6/14)*Entropy (Strong)

➡ Entropy(Weak) = - (6/8)*log2(6/8) - (2/8)*log2(2/8) = 0.811

➡ Entropy(Strong) = - (3/6)*log2(3/6) - (3/6)*log2(3/6) = 1.00

➡ So...0.940 - (8/14)*0.811 - (6/14)*1.00 = 0.048

Advantages and Disadvantages of ID3

Advantages

➡ Using ID3 understandable prediction rules are created from the training data.

➡ ID3 Builds the fastest tree.

➡ ID3 Builds a short tree.

➡ In ID3 Only there is a need to test enough attributes until all data is classified.

➡ Using ID3 Finding of leaf nodes enables test data to be pruned, thus reducing number of tests.

Disadvantages

➡ Data may be over-fitted or over-classified, if a small sample is tested.

➡ Only one attribute at a time is tested for making a decision.

➡ Classifying continuous data may be computationally expensive, as many trees must be generated to see where to break the continuum.

8.9.3.2. C4.5 Algorithm

C4.5 is an algorithm used to generate a decision tree developed by Ross Quinlan. C4.5 is an extension of Quinlan's earlier ID3 algorithm. The decision trees generated by C4.5 can be used for classification, and for this reason, C4.5 is often referred to as a statistical classifier.

C4.5 builds decision trees from a set of training data in the same way as **ID3**, using the concept of **information entropy**. The training data is a set $S = s_1, s_2, ...$ of already classified samples. Each sample $s_i = x_1, x_2, ...$ is a vector where $x_1, x_2, ...$ represent attributes or features of the sample. The training data is augmented with a vector $C = c_1, c_2, ...$ where $c_1, c_2, ...$ represent the class to which each sample belongs.

At each node of the tree, C4.5 chooses one attribute of the data that most effectively splits its set of samples into subsets enriched in one class or the other. Its criterion is the normalized **information gain** (difference in entropy) that results from choosing an attribute for splitting the data. The attribute with the highest normalized information gain is chosen to make the decision. The C4.5 algorithm then recurses on the smaller sublists.

This algorithm has a few base cases.

- All the samples in the list belong to the same class. When this happens, it simply creates a leaf node for the decision tree saying to choose that class.
- None of the features provide any information gain. In this case, C4.5 creates a decision node higher up the tree using the expected value of the class.
- Instance of previously-unseen class encountered. Again, C4.5 creates a decision node higher up the tree using the expected value.

Pseudocode

In **pseudocode**, the general algorithm for building decision trees is:

1. Check for base cases
2. For each attribute *a*

   Find the normalized information gain from splitting on *a*
3. Let *a_best* be the attribute with the highest normalized information gain
4. Create a decision *node* that splits on *a_best*
5. Recurse on the sublists obtained by splitting on *a_best*, and add those nodes as children of *node*

Improvements from ID3 Algorithm

C4.5 made a number of improvements to ID3. Some of these are:

- Handling both continuous and discrete attributes: In order to handle continuous attributes, C4.5 creates a threshold and then splits the list into those whose attribute value is above the threshold and those that are less than or equal to it.
- Handling training data with missing attribute values: C4.5 allows attribute values to be marked as? for missing. Missing attribute values are simply not used in gain and entropy calculations.
- Handling attributes with differing costs.
- Pruning trees after creation: C4.5 goes back through the tree once it's been created and attempts to remove branches that do not help by replacing them with leaf nodes. There are 2 primary pruning strategies proposed in C4.5
  (*a*) *Subtree replacement:* In this the subtree is replaced by a leaf node if this replacement results in an error rate clse to that of original tree. Subtree replacement works for bottom up approach.
  (*b*) *Subtree raising:* This method replaces a subtree by its most used subtree. In this the subtree is raised from its current location to a node higher up in the tree. Then we have to calculate the error rate for this replacement.

Gain ration in C4.5

$$\text{Gain Ratio } (D, S) = \frac{Gain\,(D, S)}{H\left(\frac{|D_1|}{|D|}, \ldots, \frac{|D_s|}{|D|}\right)}$$

8.9.3.3. CART: Classification and Regression Trees

CART is a decision tree algorithm for both classification and regression. It was first described by [Breiman1984]. It is a recursive algorithm, which partitions the training dataset by doing binary splits. It is a conceptual simple decision tree algorithm, and performs OK in practice. Like ID3 entropy is used as a measure to choose the best splitting attribute and crietrian. In this 2 children are created instead one child as in ID3. The splitting is performes around what is determined to be the best split point.

Formula for gain ration

$$\Phi(s/t) = 2P_L P_R \sum_{j=1}^{m} |P(C_j \mid t_L) - P(C_j \mid t_R)|$$

Where L and R are used to indicate left and right subtrees of the current node in the tree. P_L, P_R probability that a tuple in the training set will be on the left or right side of the tree, t is the current node, s is the splitting attribute and crietrian. This is defined as tuples in subtree/tuples in the training set. At each step, only one criterian is chosen as the best over all possible criteria.

8.10. GENETIC ALGORITHM

8.10.1. Introduction to Genetic Algorithms (GA)

Genetic algorithms (GA) are an evolutionary optimization approach, which are an alternative to traditional optimization methods. GA is one of the most appropriate methods for complex non-linear models where location of the global optimum is a difficult task. The basis for evolutionary computing algorithms is biological evolution, where evolution produces the best or fittest individuals over time.

8.10.2. Working Principle of GA

The GA procedure is based on the Darwinian principle of survival of the fittest.

- An initial population is created containing a predefined number of individuals, each represented by a genetic string.
- Each individual has an associated fitness measure. The concept that the fittest (or best) individuals in a population will produce a fitter offspring is then implemented in order to reproduce the next population.
- Selected individuals are chosen for reproduction (or crossover) at each generation, with an appropriate mutation factor to randomly modify the genes of an individual, in order to develop the new population.
- The result is another set of individuals based on the original subjects leading to subsequent populations with better fitness and those with lower fitness will naturally get discarded from the population.

The GA consists of four main stages: *evaluation*, *selection*, *crossover* and *mutation*.

- The ***evaluation*** procedure measures the fitness of each individual solution in the population and assigns to it a score.
- The ***selection*** procedure randomly selects individuals of the current population for development of the next generation.
- The ***crossover*** procedure takes two selected individuals and combines them about a crossover point thereby creating two new individuals.
- The ***mutation*** procedure randomly modifies the genes of an individual subject to a small mutation factor, introducing further randomness into the population. This iterative process continues until

one of the possible termination criteria is met: if a known optimal or acceptable solution level is attained; or if a maximum number of generations have been performed; or if a given number of generations without fitness improvement occur

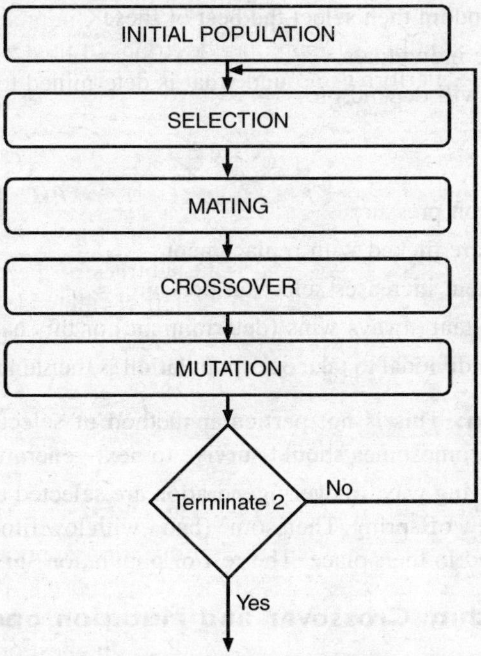

Fig. 8.13. Simplified flow chart of a Genetic Algorithm.

Although individuals are represented as bit strings, any encoding is possible. **Binary encoding** is the most common one. Other types of genetic Encoding are

Permutation encoding can be used in ordering problems

Direct value encoding can be used in problems where some more complicated values are required

Tree encoding is used mainly for evolving programs or expressions (*i.e.*, genetic programming)

8.10.3. Genetic Algorithms: Selection Operators

According to Darwin's theory of evolution the best chromosome survive to create new offspring. There are many methods in selecting the best chromosomes:

1. Roulette wheel selection: This can be simulated by the following algorithm:

1. [Sum] Calculate sum S of all chromosome fitness values in population.
2. [Select] Generate random number r from interval (0, S).
3. [Loop] Go through the population and sum fitness values from 0 - S. When the sum is greater then r, stop and return the chromosome where you are.

2. Rank selection: The previous selection will have problems when the fitness values differ very much. For example, if the best chromosome fitness is 90% of all the roulette wheel then the other chromosomes will have very few chances to be selected.

Rank selection first ranks the population and then every chromosome receives fitness from this ranking. The worst will have fitness 1, second worst 2 etc. and the best will have fitness N (number of chromosomes in population).

After this all the chromosomes have a chance to be selected. But this method can lead to slower convergence, because the best chromosomes do not differ so much from other ones.

Tournament selection Tournament selection is an informal Procedure:

- Pick k members at random then select the best of these
- Repeat to select more individuals

Probability of selecting i will depend on:

- Rank of i
- Size of sample k

Higher k increases selection pressure

- Whether contestants are picked with replacement

Picking without replacement increases selection pressure

- Whether fittest contestant always wins (deterministic) or this happens with probability p

For $k = 2$, time for fittest individual to take over population is the same as linear ranking with $s = 2 \bullet p$

3. Steady-State Selection: This is not particular method of selecting parents. Main idea of this selection is that big part of chromosomes should survive to next generation.

GA then works in a following way. In every generation are selected a few (good - with high fitness) chromosomes for creating a new offspring. Then some (bad - with low fitness) chromosomes are removed and the new offspring is placed in their place. The rest of population survives to new generation

8.10.4. Genetic Algorithm: Crossover and Mutation operators

There are many methods for crossovers and mutations depending on the kind of encoding applied to represent the chromosomes.

- **Single point crossover:** One crossover point is selected, binary string from the beginning of the chromosome to the crossover point is copied from the first parent, the rest is copied from the other parent

  11001011+11011111 = 11001111

- **Two point crossover:** Two crossover points are selected, binary string from the beginning of the chromosome to the first crossover point is copied from the first parent, the part from the first to the second crossover point is copied from the other parent and the rest is copied from the first parent again.

  11001011 + 11011111 = 11011111

- **Uniform crossover:** Bits are randomly copied from the first or from the second parent

  11001011 + 11011101 = 11011111

- Arithmetic crossover: some arithmetic operation is performed to make a new offspring (*e.g.*, logic AND)

  11001011 + 11011111 = 11001001 (AND)

Mutation: inversion of selected bits

11001001 => 10001001

GA algorithms are used to solve most data mining problems including classification, clustering, and generating the association rules. Typical applications of GA include scheduling, robotics, biology, economics etc.

8.10.5. Advantages of Limitations of GA

Advantages

- GAs work with a coding for the parameter set, not the parameters themselves.
- GAs search from a population of points, not a single point.
- GAs use probabilistic transition rules, not deterministic rules.
- The Genetic Algorithms are first suited for the optimization problem.
- GAs are robust with respect to local minima, maxima.
- GAs work well on mixed discrete/continuous problems.
- GAs can operate on various representations.
- GAs are stochastic.
- GAs are easily parallelised.
- GAs are easy to implement and lend themselves well to hybridization.

Limitations

- GA implementation is still an art. They often require a lot of modification, then again sometimes you can modify all you want and they will still find the same result.
- GAs are computationally expensive.
- GAs typically require less information about the problem (*e.g.*, no gradients), but designing an objective function and getting the representation and operators right can be difficult.
- Determining fitness function is difficult.
- Determining how to perform crossover and mutation is difficult.

8.11. ROUGH SETS

A **rough set**, is a formal approximation of a crisp set (*i.e.*, conventional set) in terms of a pair of sets which give the *lower* and the *upper* approximation of the original set. In the standard version of rough set theory (Pawlak 1991), the lower- and upper-approximation sets are crisp sets, but in other variations, the approximating sets may be fuzzy sets. Rough sets constitute a sound basis for KDD. It offers mathematical tools to discover patterns hidden in data

There are some issues relating with real world data set like Real world data always contains

1. mixed types of data such as continuous valued, symbolic data,
2. Very large data sets
3. Uncertainty (noisy data)
4. Incompleteness (missing, incomplete data)
5. Data change
6. Use of background knowledge etc.

Rough sets can be applied only to discrete valued attributes.

When it comes to analyze attributes with real values, they must undergo a process called discretization, which divides the attribute's value into intervals. There is a lack of the unified approach to discretization problems so far, and the choice of method depends heavily on data considered.

Rough sets are used to approximately or "roughly" define equivalent classes within the given training data. A rough set for a given class C is approximated by two sets:

1. a lower approximation (certain to be in C) : the lower approximation of C consists of all the data samples that based on the knowledge of the attributes are belong to C without an ambiguity.
2. an upper approximation (cannot be described as not belonging to C) : The upper approximation of C consists of all the samples that based on the knowledge of the attributes cannot be described as not belonging to Class C.

The problem of finding the minimal subsets of attributes (for feature reduction) is NP-hard

8.11.1. A Rough Set Based KDD Process

- ➡ Discretization based on RS and Boolean Reasoning (RSBR).
- ➡ Attribute selection based RS with Heuristics (RSH).
- ➡ Rule discovery by GDT-RS. (GDT : Generalization Distribution Table and RS : Rough Sets)

AS already said that rough sets work on discrete data and if data is not discrete then it must undergo discretization process. There is a lack of the unified approach to discretization problems so far, and the choice of method depends heavily on data considered.

In the discretization process, we search for a set of cuts satisfying some natural conditions.

U	a	b	d
x1	0.8	2	1
x2	1	0.5	0
x3	1.3	3	0
x4	1.4	1	1
x5	1.4	2	0
x6	1.6	3	1
x7	1.3	1	1

P = {(a, 0.9), (a, 1.5), (b, 0.75), (b, 1.5)}

U	a^P	b^P	d
x1	0	2	1
x2	1	0	0
x3	1	2	0
x4	1	1	1
x5	1	2	0
x6	2	2	1
x7	1	1	1

Fig. 8.14. Rough Set Based KDD Process.

The values smaller than 0.9 in column a will become 0, the values between 0.9 and 1.5 are represented by 1 and values greater than 1.5 is represented as 2 similarly for b.

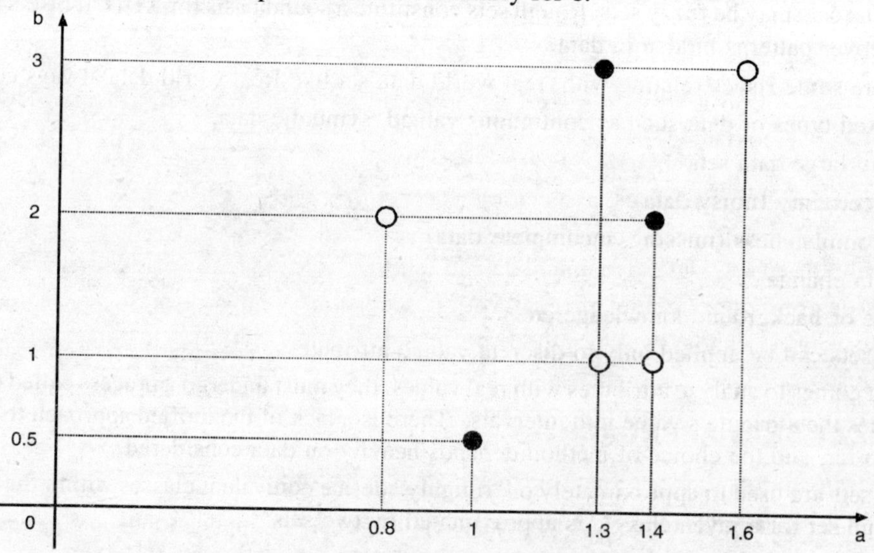

Fig. 8.15. A Geometrical Representation of Data.

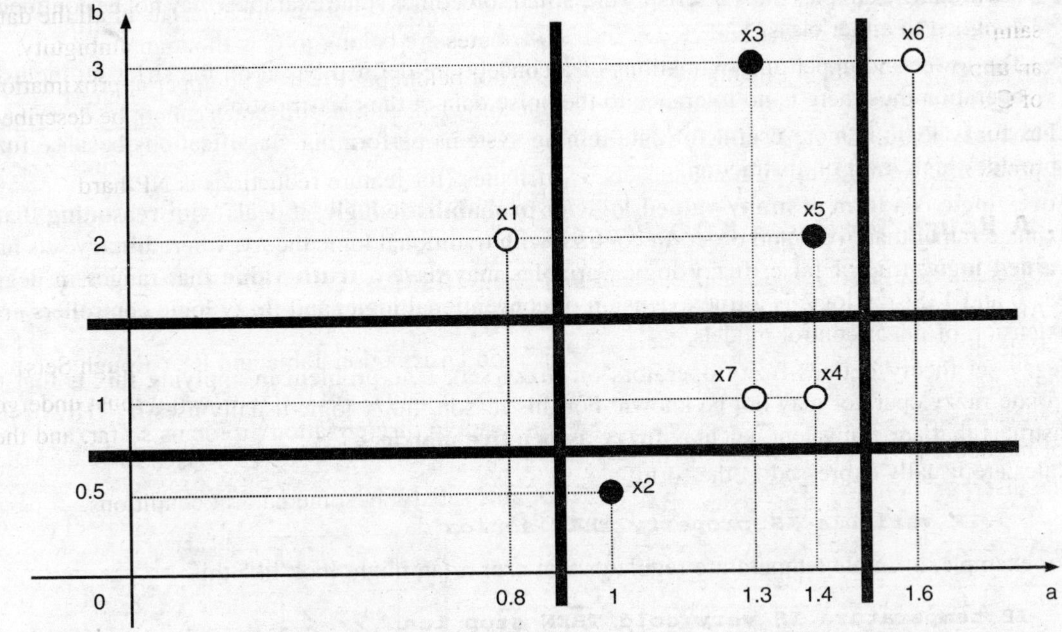

Fig. 8.16. Geometrical Representation of Data and Cuts (Cuts are Shown by Dark Lines).

The discretization process returns a partition of the value sets of condition attributes into intervals.

Rough Sets Application

Rough set methods can be applied as a component of hybrid solutions in machine learning and data mining. They have been found to be particularly useful for rule induction and feature selection (semantics-preserving dimensionality reduction). Rough set-based data analysis methods have been successfully applied in bioinformatics, economics and finance, medicine, multimedia, web and text mining, signal and image processing, software engineering, robotics, and engineering (*e.g.*, power systems and control engineering).

It can be used for

1. feature selection,
2. feature extraction,
3. data reduction,
4. decision rule generation, and
5. pattern extraction (templates, association rules) etc.
6. It identifies partial or total dependencies in data, eliminates redundant data, gives approach to null values, missing data, dynamic data and others.

Recent extensions of rough set theory (**rough mereology**) have developed new methods for decomposition of large data sets, data mining in distributed and multi-agent systems, and granular computing.

8.12. FUZZY SETS APPROACH

There are some limitations of Rough Set Theory like:

➡ All tuples are treated with equal importance but In real-life application, some *typical* examples may be more important than others.

- All training examples must be crisply classified sometimes some examples may not be confidently assigned to either class.
- The lower and upper approximations of a concept are defined based on the strict set inclusion operation thus there is no tolerance to the noise data in the classification.

Thus fuzzy logic is more useful for data mining systems performing classifications because fuzzy logic provides high level of abstraction.

Fuzzy logic is a form of **many-valued logic** or **probabilistic logic**; it deals with **reasoning** that is approximate rather than fixed and exact. In contrast with traditional logic theory, where **binary** sets have **two-valued logic**: true or false, fuzzy logic variables may have a **truth value** that ranges in degree between 0 and 1. Fuzzy logic is a true extension of conventional logic, and fuzzy logic controllers are a true extension of linear control models.

Fuzzy set theory defines fuzzy operators on fuzzy sets. The problem in applying this is that the appropriate fuzzy operator may not be known. For this reason, fuzzy logic usually uses IF-THEN rules, or constructs that are equivalent, such as **fuzzy associative matrices**.

Rules are usually expressed in the form:

```
IF variable IS property THEN action
```

For example, a simple temperature regulator that uses a fan might look like this:

```
IF temperature IS very cold THEN stop fan
IF temperature IS cold THEN turn down fan
IF temperature IS normal THEN maintain level
IF temperature IS hot THEN speed up fan
```

There is no "ELSE" – all of the rules are evaluated, because the temperature might be "cold" and "normal" at the same time to different degrees.

8.12.1. Advantages and Disadvantages of Fuzzy Logic

Advantages

1. Fuzzy logic allows the use of vague linguistic terms in the rules.
2. Fuzzy logic Can accept multiple inputs and generate multiple results.
3. For fuzzy logic no complex mathematical model is required.

Disadvantages

1. Since in Fuzzy logic each element is associated with its membership function and it is difficult to estimate membership function
2. There are many ways of interpreting fuzzy rules, combining the outputs of several fuzzy rules and defuzzification output.

8.13. SUPPORT VICTOR MACHINE [MDU Dec. 2011]

A **support vector machine** (SVM) is a concept in **statistics** and **computer science** for a set of related **supervised learning** methods that analyze data and recognize patterns, used for **classification** and **regression analysis**.

The standard SVM takes a set of input data and predicts, for each given input, which of two possible classes forms the input, making the SVM a non-**probabilistic binary linear classifier**.

Given a set of training examples, each marked as belonging to one of two categories, an SVM training algorithm builds a model that assigns new examples into one category or the other. An SVM model is a representation of the examples as points in space, mapped so that the examples of the separate categories are divided by a clear gap that is as wide as possible. New examples are then mapped into that same space and predicted to belong to a category based on which side of the gap they fall on.

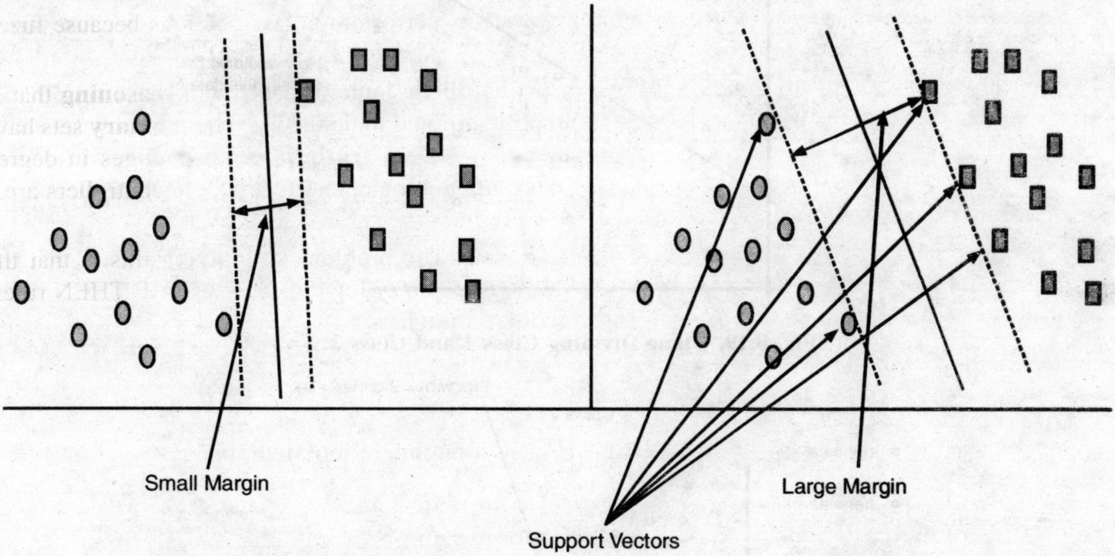

Fig. 8.17. Small Margin Vs. Large Margin in Support Vectors.

Definition: A support vector machine constructs a hyperplane or set of hyperplanes in a high- or infinite-dimensional space, which can be used for classification, regression, or other tasks. Intuitively, a good separation is achieved by the hyperplane that has the largest distance to the nearest training data point of any class (so-called functional margin), since in general the larger the margin the lower the generalization error of the classifier.

SVM-General Concept

We have already studied the concept of linear classifier in ANN.

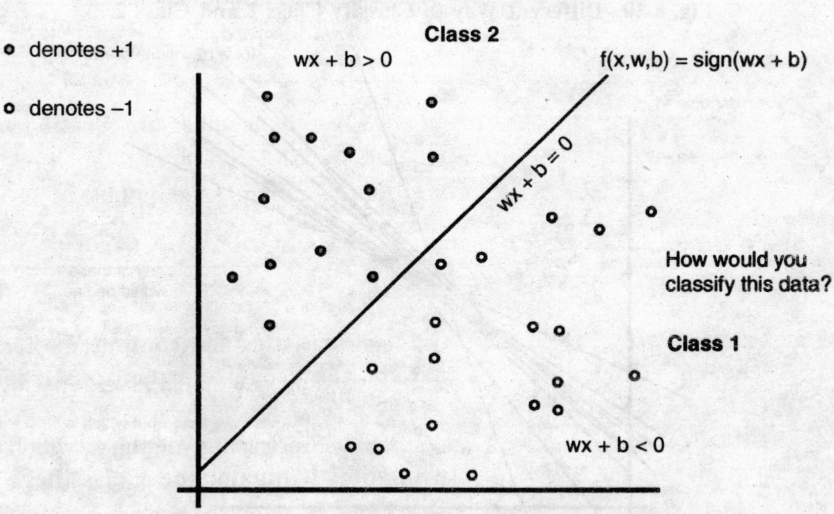

Fig. 8.18. Plane Dividing Class 1 and Class 2.

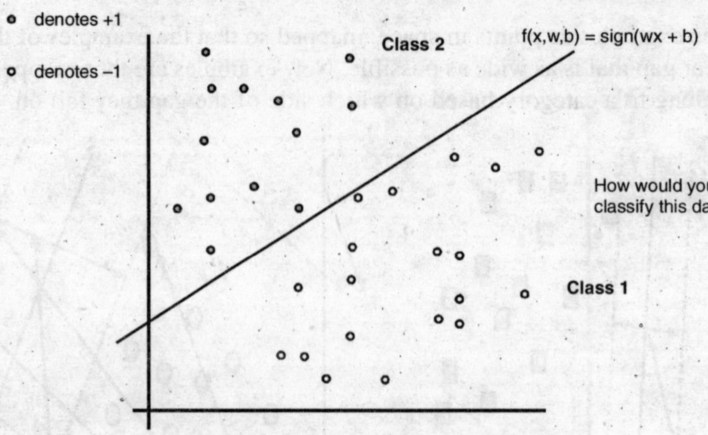

Fig. 8.19. Plane Dividing Class 1 and Class 2.

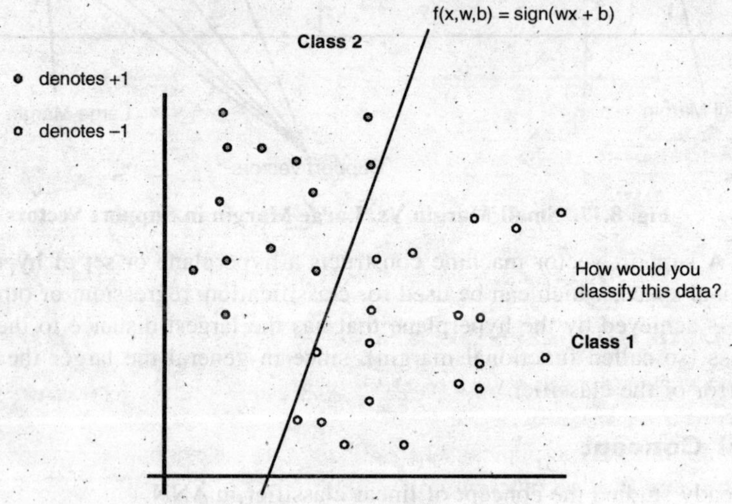

Fig. 8.20. Different Way to Classify Class 1 and Class 2.

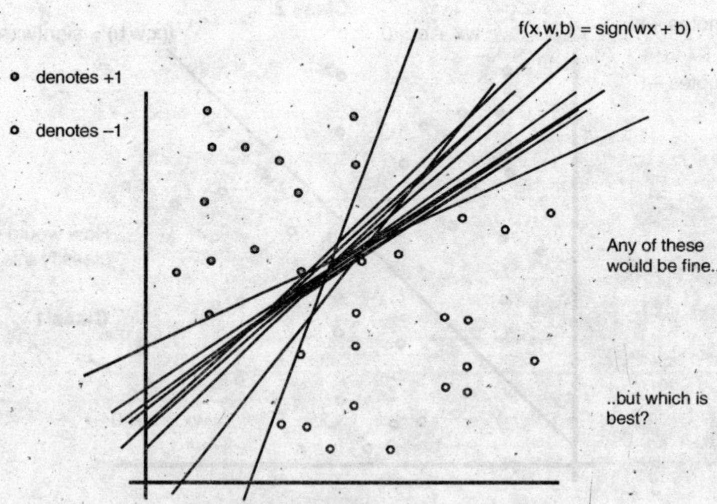

Fig. 8.21.

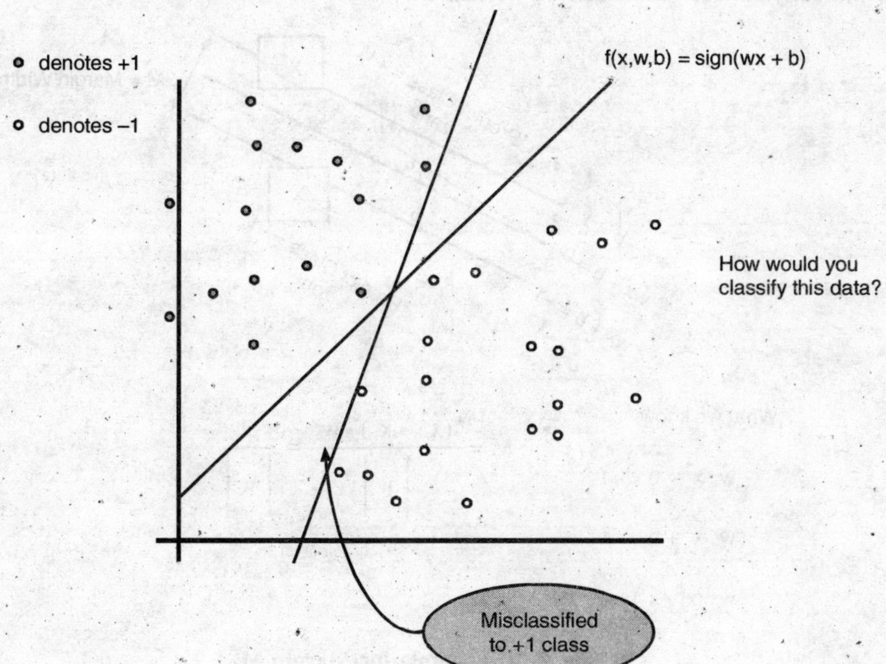

Fig. 8.22.

SVM—When Data Is Linearly Separable

Let data D be $(\mathbf{X}_1, y_1), \ldots, (\mathbf{X}_{|D|}, y_{|D|})$, where $\mathbf{X}_i$ is the set of training tuples associated with the class labels y_i.

There are infinite lines (hyperplanes) separating the two classes but we want to find the best one (the one that minimizes classification error on unseen data)

SVM searches for the hyperplane with the largest margin, *i.e.*, maximum marginal hyperplane (MMH).

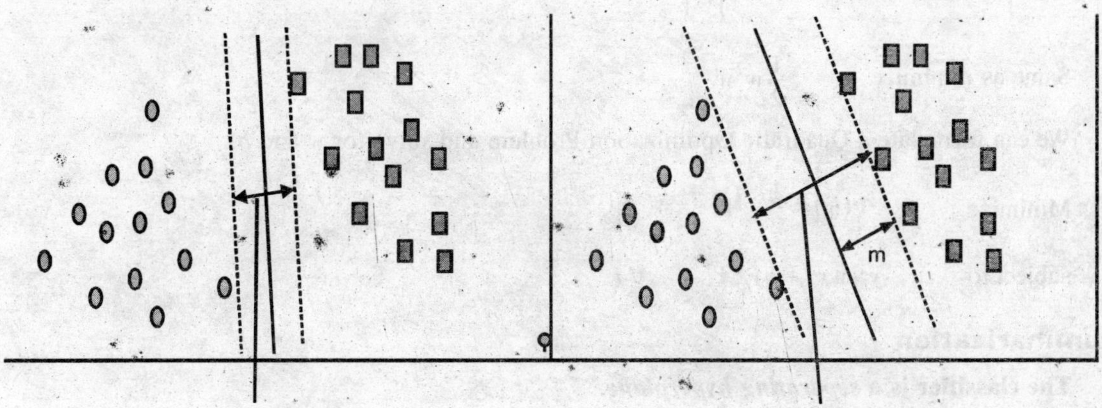

Fig. 8.23. Small Vs Large Margin

Mathematically Linear Vector can be written as

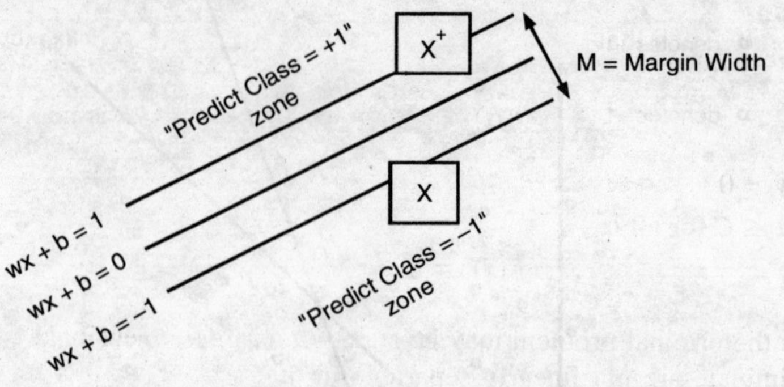

What we know:

- $w \cdot x^+ + b = +1$
- $w \cdot x^- + b = -1$
- $w \cdot (x^+ - x^-) = 2$

$$M = \frac{(x^+ - x^-) \cdot w}{|w|} = \frac{2}{|w|}$$

Fig. 8.24. Formula for Margin M.

Any training tuples that fall on hyperplanes H_1 or H_2 (*i.e.*, the sides defining the margin) are **support vectors.**

Goal: 1. Correctly classify all training data

$$\left.\begin{aligned} wx_i + b \geq 1 \quad & \text{if } y_i = +1 \\ wx_i + b \leq 1 \quad & \text{if } y_i = -1 \end{aligned}\right\}$$

$$y_i(wx_i + b) \geq 1 \quad \text{for all } i$$

(2) Maximum the Margin $M = \dfrac{2}{|w|}$

Same as minimize $\dfrac{1}{2} w^t w$

We can formulate a Quadratic Optimization Problem and solve for w and b.

Minimize $\Phi(w) = \dfrac{1}{2} w^t w$

subject to $y_i(wx_i + b) \geq 1 \quad \forall i$

Summarization

The classifier is a *separating hyperplane.*

- Most "important" training points are support vectors; they define the hyperplane.
- Quadratic optimization algorithms can identify which training points x_i are support vectors with non-zero Lagrangian multipliers $\acute{a}_i$.

Both in the dual formulation of the problem and in the solution training points appear only inside dot products:

Find $\alpha_1 \ldots \alpha_N$ such that

$$Q(\alpha) = \Sigma\alpha_i - \frac{1}{2}\Sigma\Sigma\alpha_i\alpha_j y_i y_j x_i^T x_j \text{ is maximized and}$$

(1) $\Sigma\alpha_i y_i = 0$

(2) $0 \le \alpha_i \le C$ for all α_i.

$$f(x) = \Sigma\alpha_i y_i x_i^T x + b$$

Whereas the original problem may be stated in a finite dimensional space, it often happens that the sets to discriminate are not linearly separable in that space. For this reason, it was proposed that the original finite-dimensional space be mapped into a much higher-dimensional space.

To keep the computational load reasonable, the mappings used by SVM schemes are designed to ensure that dot products may be computed easily in terms of the variables in the original space, by defining them in terms of a kernel function $K(Xi, Xj)$ selected to suit the problem

i.e., $$K(\mathbf{X_i}, \mathbf{X_j}) = \Phi(\mathbf{X_i})\ \Phi(\mathbf{X_j})$$

Typical Kernel Functions

Polynomial kernal of degree h : $K(\mathbf{X}_i, \mathbf{X}_j) = (\mathbf{X}_i \,.\, \mathbf{X}_j + 1)^n$

Gaussian radial basis function kernel : $K(\mathbf{X}_i, \mathbf{X}_j) = e^{-\|\mathbf{X}_i \,.\, \mathbf{X}_j\|^2/2\sigma^2}$

Sigmoid kernel : $K(\mathbf{X}_i, \mathbf{X}_j) = \tanh(\kappa\mathbf{X}_i \,.\, \mathbf{X}_j - \delta)$

8.13.1. Properties of SVM

- Flexibility in choosing a similarity function.
- Sparseness of solution when dealing with large data sets
- only support vectors are used to specify the separating hyperplane
- Ability to handle large feature spaces
- complexity does not depend on the dimensionality of the feature space
- Overfitting can be controlled by soft margin approach
- Nice math property: A simple convex optimization problem which is guaranteed to converge to a single global solution.

Application of SVM

SVM used for:

1. Handwritten digit recognition,
2. Text (and hypertext) categorization
3. Image classification
4. Bioinformatics (Protein classification, Cancer classification)
5. Hand-written character recognition
6. Time-series prediction tests

Drawbacks of SVM

Potential drawbacks of the SVM are the following three aspects:

- Uncalibrated class membership probabilities
- The SVM is only directly applicable for two-class tasks. Therefore, algorithms that reduce the multi-class task to several binary problems have to be applied; see the multi-class SVM section.
- Parameters of a solved model are difficult to interpret

Summary

- **Classification vs. Prediction:** Classification: predicts categorical class labels and classifies data (constructs a model) based on the training set and the values (class labels) in a classifying attribute and ....*uses it in classifying new data*

  Prediction: models continuous-valued functions, *i.e.,* predicts *unknown or missing values*
- **Application of Classification and Prediction** implies

  credit approval

  target marketing

  medical diagnosis

  treatment effectiveness analysis
- **Supervised vs. Unsupervised Learning:** Supervised learning *(classification)* : The training data (observations, measurements, etc.) are accompanied by labels indicating the class of the observations and New data is classified based on the training set.

  Unsupervised learning (clustering): The class labels of training data is unknown. Given a set of measurements, observations, etc. with the aim of establishing the existence of classes or clusters in the data.
- Decision tree generation consists of two phases

  1. Tree construction: (*a*) At start, all the training examples are at the root.

  (*b*) Partition examples recursively based on selected attributes

  2. Tree pruning: Identify and remove branches that reflect noise or outliers
- Id3, C4.5 and CART are the decision tree algorithms based on entropy and information theory.
- Naïve Bayesian classification is based on Bayes theorem. Naïve Bayesian classification assumes class conditional independencies but Bayesian belief networks allow class independencies to be defines between subset of variables.
- Backpropagation is a supervised feedforwad NN algorithm used for classification and it employs gradient descent. Rules may be extracted from trained NN in order to help improve the interpretability of the learned network.
- **GA:** based on an analogy to biological evolution and Each rule is represented by a string of bits. In GA An initial population is created consisting of randomly generated rules

  For example, IF A_1 and Not A_2 then C_2 can be encoded as 100. The fitness of a rule is represented by its classification accuracy on a set of training examples

  Offsprings are generated by *crossover and mutation.*
- Rough sets are used to approximately or "roughly" define equivalent classes. A rough set for a given class C is approximated by two sets:
  - a lower approximation (certain to be in C) and
  - an upper approximation (cannot be described as not belonging to C).

- Fuzzy set theory is another approach used for classification purpose. Fuzzy logic uses truth values between 0.0 and 1.0 to represent the degree of membership (such as using fuzzy membership graph.
- **Difference between SVM and ANN**

SVM	Neural Network
1. Deterministic algorithm	Nondeterministic algorithm
2. Nice generalization properties.	Generalizes well but doesn't have strong mathematical foundation
3. Hard to learn—learned in batch mode using quadratic programming techniques.	Can easily be learned in incremental fashion.
4. Using kernels can learn very complex functions.	To learn complex functions—use multilayer perceptron (nontrivial)

- **Soft techniques for KDD**
  1. Logic (*Deduction, Induction, Abduction*)
  2. Set (Rough Sets, Fuzzy Sets)
  3. Probability (*Stoch. Proc., Belief Nets, Conn. Nets, GDT*)
- Explain C4.5
  - ➡ Simple depth-first construction.
  - ➡ Uses Information Gain
  - ➡ Sorts Continuous Attributes at each node.
  - ➡ Needs entire data to fit in memory.
  - ➡ Unsuitable for Large Datasets.
- **Which Technique and When?:** Clearly one of the hardest things to do when deciding to implement a data mining system is to determine which technique to use when. When are neural networks appropriate and when are decision trees appropriate? When is data mining appropriate at all as opposed to just working with relational databases and reporting? When would just using OLAP and a multidimensional database be appropriate?

  Some of the criteria that are important in determining the technique to be used are determined by trial and error. There are definite differences in the types of problems that are most conducive to each technique but the reality of real world data and the dynamic way in which markets, customers and hence the data that represents them is formed means that the data is constantly changing. These dynamics mean that it no longer makes sense to build the "perfect" model on the historical data since whatever was known in the past cannot adequately predict the future because the future is so unlike what has gone before.

  In some ways this situation is analogous to the business person who is waiting for all information to come in before they make their decision. They are trying out different scenarios, different formulae and researching new sources of information. But this is a task that will never be accomplished - at least in part because the business the economy and even the world is changing in unpredictable and even chaotic ways that could never be adequately predicted. Better to take a robust model that perhaps is an under-performer compared to what some of the best data mining tools could provide with a great deal of analysis and execute it today rather than to wait until tomorrow when it may be too late.

Review Questions

1. **Explain the terms classification and predication.**
2. **What do u mean by supervised and unsupervised learning.**
3. **How Decision trees are used for classification.**
4. **Explain the term tree pruning and its types.**
5. **Write the advantages and disadvantages of ID3, C 4.5, CART.**
6. **Derive the expression for probability in NAÏVE BAYESIAN classification?**
7. **How Backpropagation is used for classification.**
8. **Write a short note on Genetic Algorithm, rough set and fuzzy set theory.**
9. **Explain Overfitting and the method to overcome it.**
10. **Is there any advantages of Fuzzy logic over rough set theory, explain.**

❑❑❑

Chapter 9

Association Rules

INSIDE THIS CHAPTER

9.1. Introduction **9.2.** Concept of Frequent Itemsets, Closed Itemsets and Association Rules **9.3.** Rule Strength Measures **9.4.** Classification of Association Rules **9.5.** Basic Algorithms For Mining Single Dimensional Boolean Association Rules From Transactional Databases **9.6.** Iceberg Queries **9.7.** Different Approaches For Mining Multilevel Association Rule **9.8.** Mining Multi Dimensional Association Rules from Relational Databases and Data Warehouse **9.9.** Basic Approaches for Mining Multidimensional Association Rules

This chapter deals with association rules mining and various algorithms for Mining single-dimensional Boolean association rules from transactional databases: Apriori Algorithm, Sampling Algorithm, Partitioning, Mining multilevel-dimensional association rules from transactional database and Mining multilevel-dimensional association rules from relational databases

9.1. INTRODUCTION

Association rule learning is a popular and well researched method for discovering interesting relations between variables in large databases. Lets understand the concept by looking at the customer buying behavior in a store. An example rule for the store could be

$$\{\text{teabags, sugar}\} \Rightarrow \{\text{milk, snacks}\}$$

meaning that if teabags and sugar are bought, customers are also likely to buy milk and snacks.

Association rules are employed today in many application areas including Web usage mining, intrusion detection and bioinformatics. As opposed to sequence mining, association rule learning typically does not consider the order of items either within a transaction or across transactions

In data mining, it is a method for discovering relations between different sets of items in a large database and Database is A large collection of transactions, Example - Market basket database. The definition include

$$X \Rightarrow Y$$

$$X = \{x_1, x_2, ..., x_n\}$$
$$Y = \{y_1, y_2, ..., y_n\}$$

x_i and y_j are distinct items for all i and all j.
X is the left-hand-side (LHS) and Y is the right-hand-side (RHS).

9.2. CONCEPT OF FREQUENT ITEMSETS, CLOSED ITEMSETS AND ASSOCIATION RULES

Frequent Itemsets

1. A set of items is referred to as an itemset. An itemset that contains *k* items is a *k*-itemset. The set {computer, antivirus software}is a 2-itemset.
2. The occurrence frequency (called support count, frequency or count)of an itemset is the number of transactions that contain the itemset.
3. If the relative support of an itemset I satisfies a prespecified minimum support threshold (*i.e.*, the absolute support of I satisfies the corresponding minimum support count threshold), then I is a frequent itemset.
4. The set of frequent *k*-itemsets is commonly denoted by L_k.

Closed Itemsets

An itemset *X* is closed in a data set *S* if there exists no proper super-itemset *Y* such that *Y* has the same support count as *X* in *S*. An itemset *X* is a closed frequent itemset in set *S* if *X* is both closed and frequent in *S*. An itemset *X* is a maximal frequent itemset (or max-itemset) in set *S* if *X* is frequent, and there exists no super-itemset *Y* such that *XY* and *Y* is frequent in *S*.

Association Rules

1. They are used for finding patterns or relationship in the database.
2. For example, the information that customers who purchase computers also tend to buy antivirus software at the same time is represented in Association Rule below:

   Computer → *antivirus_software [support = 2%; confidence = 60%]*

2. Rule support and confidence are two measures of rule interestingness. They respectively reflect the usefulness and certainty of discovered rules.
3. A support of 2% for Association Rule means that 2% of all the transactions under analysis show that computer and antivirus software are purchased together.
4. A confidence of 60% means that 60% of the customers who purchased a computer also bought the software.
5. Typically, association rules are considered interesting if they satisfy both a minimum support threshold and a minimum confidence threshold.
6. Such thresholds can be set by users or domain experts. Additional analysis can be performed to uncover interesting statistical correlations between associated items.

9.3. RULE STRENGTH MEASURES

Rule strength measures are:

(*i*) Support
(*ii*) Confidence

9.3.1. Support

The support of an itemset expresses how often the itemset appears in a single transaction in the database, *i.e.*, the support of an item is the percentage of transactions in which that item (or items) occurs.

For an association rule, $A \rightarrow B$

Support(A → B) = number of tuples containing both A and B / total number of tuples

For example,

ITEMSET={teabags, sugar, milk, surf}

Example database with 4 items and 5 transactions

Transaction ID	teabags	sugar	milk	surf
1	1	1	0	0
2	0	0	1	0
3	0	0	0	1
4	1	1	1	0
5	0	1	0	0

In the example database, the itemset ={teabags, sugar, milk, surf} has a support of 1/5 = 0.2 since it occurs in 20% of all transactions (1 out of 5 transactions).

Confidence: The confidence or strength for an association rule $A \rightarrow B$ is the ratio of the number of transaction that contain $A \rightarrow B$ to the number of transactions that contain A.

Confidence is defined as the measure of certainty associated with each discovered pattern. For an association rule, $A \rightarrow B$

Confidence($A \rightarrow B$) = no. of tuples containing both A and B /No of tuples containing A.

Thus,conf($X \Rightarrow Y$) = supp ($X \Rightarrow Y$)/supp(X)

As from above transaction table, the rule {teabags, sugar}à milk has a confidence of 0.2/0.4 = 50% or 0.5 in the database.

supp($X \cap$ Y) means "*support for occurrences of transactions where* ***X and Y both appear***", not "*support for occurrences of transactions where* ***either X or Y appears.***

Another Example

Transaction ID	Items Bought
1	Milk, bread, cookies, juice
2	Milk, juice
3	Milk, eggs
4	Bread, cookies, coffee

Rules:

- Milk => juice
- Bread => juice

Support {milk, juice} is 2 / 4 = 0.50 and {bread, juice}= 1 / 4 = 0.25

- Confidence Milk => juice, 0.50 / 0.75 = 0.67
- Bread => juice, 0.25 / 0.50 = 0.50

9.4. CLASSIFICATION OF ASSOCIATION RULES

Association rules can be classified in various ways, discussed below :

1. Based on the types of values handled in the rule.
2. Based on the dimensions of data involved in the rule.
3. Based on the levels of abstraction involved in the rule.
4. Based on various extension to association mining.

9.4.1. Based on the Completeness of Patterns to be Mined

We can mine the complete set of frequent itemsets, the closed frequent itemsets, and the maximal frequent itemsets, given a minimum support threshold. We can also mine constrained frequent itemsets, approximate frequent itemsets, near-match frequent itemsets, top-*k* frequent itemsets.

Different applications may have different requirements regarding the completeness of the patterns to be mined, which in turn can lead to different evaluation and optimization methods.

9.4.2. Based on the Levels of Abstraction Involved in the Rule Set

Some methods for association rule mining can find rules at differing levels of abstraction. For example,

buys(X, "computer") → buys(X, "HP printer")
buys(X, "laptop computer") → buys(X, "HP printer")

In these Rules, the items bought are referenced at different levels of abstraction. The rule set mined is consisting of multilevel association rules. If, instead, the rules within a given set do not reference items or attributes at different levels of abstraction, then the set contains single-level association rules. E.g. only the rule

buys(X, "computer") → buys(X, "HP printer")

9.4.3. Based on the Number of Data Dimensions Involved in the Rule

If the items or attributes in an association rule reference only one dimension, then it is a single-dimensional association rule . For example,

buys(X, "computer") → buys(X, "antivirus software")

If a rule references two or more dimensions, then it is a multidimensional association rule. For example,

age(X, "30...39")^income(X, "42K...48K")?buys(X, "high resolution TV"):

9.4.4. Based on the Types of Values Handled in the Rule

If a rule involves associations between the presence or absence of items, it is a Boolean association rule. For example,

Rules buys(X, "computer")) buys(X, "HP printer") are Boolean association rules.

If a rule describes associations between quantitative items or attributes, then it is a quantitative association rule.

age(X, "30...39")^income(X, "42K...48K") → buys(X, "high resolution TV"):

Age and income, have been discretized.

9.4.5. Based on the Kinds of Rules to be Mined

Frequent pattern analysis can generate various kinds of rules and other interesting relationships. Many of the rules are redundant or do not indicate a correlation relationship among itemsets. Thus, the discovered associations can be further analyzed to uncover statistical correlations, leading to correlation rules.

We can also mine strong gradient relationships among items. Example "The average sales from Sony Digital Camera increase over 16% when sold together with Sony Laptop Computer": both Sony Digital Camera and Sony Laptop Computer are siblings, where the parent itemset is Sony.

9.4.6. Based on the Kinds of Patterns to be Mined

Many kinds of frequent patterns can be mined from different kinds of data sets like frequent itemset mining, Sequential pattern mining, Structured pattern mining.

9.5. BASIC ALGORITHMS FOR MINING SINGLE DIMENSIONAL BOOLEAN ASSOCIATION RULES FROM TRANSACTIONAL DATABASES

9.5.1. Apriori Algorithm

For mining frequent itemsets for Boolean association rules

Apriori is a classic algorithm for learning association rules and used in many commercial products. Apriori means prior to, same is the meaning here.

The name of the algorithm is based on the fact that the algorithm uses prior knowledge of frequent itemset properties.

An iterative approach known as a level-wise search is used in this algorithm where *k*-itemsets are used to explore $(k + 1)$-itemsets. And so on. To improve the efficiency of the level-wise generation of frequent itemsets, apriori property is used.

1. Apriori property states that all nonempty subsets of a frequent itemset must also be frequent.
2. The property used in the algorithm uses two-step process. Using $L_k - 1$, L_k is found for $k >= 2$. 2 step process to show how level wise search is performed consists of join step and prune step.

The join step: To find L_k, a set of candidate *k*-itemsets is generated by joining $L_k - 1$ with itself. This set of candidates is denoted C_k. In the join component, $L_k - 1$ is joined with $L_k - 1$ to generate potential candidates.

The prune step: C_k is a superset of L_k, that is, its members may or may not be frequent, but all of the frequent *k*-itemsets are included in C_k. The prune component employs the Apriori property to remove candidates that have a subset that is not frequent.

Example 9.1.

Dataset T, supp=0.5

TID	List of items
T100	1, 3, 4
T200	2, 3, 5
T300	1, 2, 3, 5
T400	2, 5

Solution.

SCAN T	Candidates	Count	Large Itemsets
1.	{1} {2} {3} {4} {5}	2 3 3 1 3	{1} {2} {3} {5} Note {4} is removed because it has the minimum count from all of other.
2.	{1, 2} {1, 3} {1, 5} {2, 3} {2, 5} {3, 5}	1 2 1 2 3 2	{1, 3} {2, 3} {2, 5} {3, 5} Others have been removed
3.	{1, 2, 3} {1, 3, 5} {2, 3, 5}	1 0 2	{2, 3, 5} Others have been removed
4.	Φ		Φ

9.5.1.1. Generating Rules From Frequent Itemsets

Once the frequent itemsets (explained above) in a database *D* have been found, it is easy to generate strong association rules from them.this can be done using conditional probability.

Confidence(A → B)=P(B/A)=support_count(A ∪ B)/support_count(A)

From this the association rules can be generated as follows:

For each frequent itemset *X*, generate all nonempty subsets of *X*.

For every nonempty subset *A* of *X*,

Let $B = X - A$

$A \rightarrow B$ is an association rule if

$$\text{Confidence}(A \rightarrow B) \geq \text{minconf},$$

$$\text{support}(A \rightarrow B) = \text{support}(A \cup B) = \text{support}(X)$$

$$\text{confidence}(A \rightarrow B) = \text{support}(A \cup B) / \text{support}(A)$$

where minconf is the minimum confidence threshold.

9.5.1.2. Variations of Apriori to Improve the Efficiency of the Original Algorithm are

1. Hash-Based Technique (Hashing Itemsets Into Corresponding Buckets): A *k*-itemset whose corresponding hasing bucket count is below the threshold cannot be frequent thus should be removed from the candidate set. Thus this technique is helpful in reducing the number of candidate *k*-itemsets specially.

2. Transaction reduction (reducing the number of transactions scanned in future iterations): A transaction that does not contain any frequent *k*-itemsets is useless in subsequent scans thus can be removed.

3. Partitioning (partitioning the data to find candidate itemsets): Any itemset that is potentially frequent with respect to Database must occur as a frequent itemset in *t* least one of the partitions.

4. Sampling (mining on a subset of the given data): In this a random sample S of the given dataset D is chosen then search is done for frequent itemsets in S instead of D. To avoid the problem of missing some of the global frequent itemsets (search is done for frequent itemsets in S instead of D) lower support threshold is used than minimum support to find the frequent itemsets.

5. Dynamic itemset counting (adding candidate itemsets at different points during a scan): In this the database is partitioned into blocks marked by start points. A new candidate itemsets can be added only when all of their subsets are estimated to be frequent.

Summary of Apriori Algorithm

Seems to be very expensive

(*i*) Level-wise search
(*ii*) K = the size of the largest itemset
(*iii*) It makes at most K passes over data
(*iv*) In practice, K is bounded (10).
(*v*) The algorithm is very fast. Under some conditions, all rules can be found in linear time.
(*vi*) Scale up to large data sets.

Example 9.2. Suppose {2,3,4} is frequent, with sup=50%

Proper nonempty subsets: {2,3}, {2,4}, {3,4}, {2}, {3}, {4}, with sup=50%, 50%, 75%, 75%, 75%, 75% respectively

These generate these association rules:

$$2, 3 \rightarrow 4, \text{ confidence} = 100\%$$
$$2, 4 \rightarrow 3, \text{ confidence} = 100\%$$
$$3, 4 \rightarrow 2, \text{ confidence} = 67\%$$
$$2 \rightarrow 3, 4, \text{ confidence} = 67\%$$
$$3 \rightarrow 2, 4, \text{ confidence} = 67\%$$
$$4 \rightarrow 2, 3, \text{ confidence} = 67\%$$

All rules have support = 50%

A Database has four transactions. Let minimum support and confidence be 50%.

Tid	Items bought
1	*A, B, D*
2	*A, D*
3	*A, C*
4	*B, D, E, F*

Find out the frequent item sets and strong association rules for the above example using Apriori Algorithm.

Solution. Step 1: Scan for support count of each candidate –{*A, B, C, D, E, F*}

C_1 :

Item	Support
A	3/4 = 75%
B	2/4 = 50%
C	1/4 = 25%
D	3/4 = 75%
E	1/4 = 25%
F	1/4 = 25%

Step 2: Comparing with the min support threshold count of 50%.

L1:

Item	Support
A	3
B	2
D	3

Step 3: Generate candidate C2 from L1.

C2 :

Item
{A,B}
{A,D}
{B,D}

Step 4: Scan for support count of each candidate in C2.

Item	Support
{A,B}	1/4 = 25%
{A,D}	2/4 = 50%
{B,D}	1/4 = 25%

Step 5: Compare C2 with minimum support

L2:

Item	Support
{A,D}	2

Step 6: Data contains frequent item {A, D}

Therefore the association rule that can be generated from L are:

Association Rule	Support	Confidence
A->D	2	2/3 = 66%
D->A	2	2/2=100%

As minimum confidence threshold is 50%, then both the rules are output as showing the confidence above 50%.

The rules are:

1. A->D   2. D->A

Example 9.3. Given minimum support as 40% and minimum confidence as 80%. Find out the frequent item sets and strong association rules for the example.

Transaction Id	Items
1	1,3,4,6
2	2,3,5,7
3	1,2,3,5,8
4	2,5,9,10
5	1,4

Solution. Step 1: Let 1 be referred to as 'I_1' and 2 as 'I_2'.

Item	Support
I_1	3/5 = 60%
I_2	3/5 = 60%
I_3	3/5 = 60%
I_4	2/5 =40%
I_5	3/5 = 60%
I_6	1/5 = 20%
I_7	1/5 = 20%
I_8	1/5 = 20%
I_9	1/5 = 20%
I_{10}	1/5 = 20%

Step 2: Minimum support threshold is 40%. Therefore,

L1:

Items	Support
I_1	3
I_2	3
I_3	3
I_4	2
I_5	3

Step 3:

C2:

Items	Support
$\{I_1, I_2\}$	1
$\{I_1, I_3\}$	2
$\{I_1, I_4\}$	2
$\{I_1, I_5\}$	1
$\{I_2, I_3\}$	2
$\{I_2, I_4\}$	0
$\{I_2, I_5\}$	3
$\{I_3, I_4\}$	1
$\{I_3, I_5\}$	2
$\{I_4, I_5\}$	0

L2:

Items	Support
$\{I_1, I_3\}$	2
$\{I_1, I_4\}$	2
$\{I_2, I_3\}$	2
$\{I_2, I_5\}$	3
$\{I_3, I_5\}$	2

C3:

Items	Support
$\{ I_1, I_2, I_3\}$	1
$\{ I_1, I_3, I_4\}$	NA
$\{ I_2, I_3, I_5\}$	2
$\{ I_1, I_2, I_3, I_5\}$	NA
$\{ I_1, I_2, I_5\}$	1
$\{ I_1, I_3, I_5\}$	1
$\{ I_2, I_3, I_5\}$	2

$\{ I_1, I_3, I_4\}$ is not to be included because $\{ I_1, I_3\}$ is a part of L2 but not $\{ I_3, I_4\}$.
$\{ I_1, I_2, I_3, I_5\}$ is not to be included because it has 4 elements.
Comparing min support count $\{ I_2, I_3, I_5\}$ is freq occurring itemset.
Rule 1: $I_3 \cap I_5 \rightarrow I_2$ Confidence = 2/2 = 100%
Rule 2: $I_2 \cap I_5 \rightarrow I_3$ Confidence = 2/3 = 66.67%
Rule 3: $I_2 \cap I_3 \rightarrow I_5$ Confidence = 2/2 = 100%
Given min confidence threshold is 80%, Rule 1 and 3 are strong association rules.

9.5.2. FP-Growth Algorithm

Only brief introduction is given here.

Algorithm: FP growth. Mine frequent itemsets using an FP-tree by pattern fragment growth.

Input:

D, a transaction database;

Min_sup, the minimum support count threshold.

Output: The complete set of frequent patterns.

Method:

1. The FP-tree is constructed in the following steps:
   (*a*) Scan the transaction database *D* once.
   Collect *F*, the set of frequent items, and their support counts.
   Sort *F* in support count descending order as *L*, the *list* of frequent items.
   (*b*) Create the root of an FP-tree, and label it as "null."
   For each transaction *Trans* in *D* do the following.
   Select and sort the frequent items in *Trans* according to the order of *L*.
   Let the sorted frequent item list in *Trans* be [*p*|*P*], where *p* is the first element and *P* is the remaining list. Call insert tree([*p*|*P*], *T*), which is performed as follows.
   If *T* has a child *N* such that *N.item-name=p.item-name*, then increment *N*'s count by 1; else create a new node *N*, and let its count be 1, its parent link be linked to *T*, and its node-link to the nodes with the same *item-name* via the node-link structure.
   If *P* is nonempty, call insert tree(*P*, *N*) recursively.
2. The FP-tree is mined by calling FP growth(*FP tree*, *null*), which is implemented as follows.
   procedure FP growth(*Tree*, α)
   (1) if *Tree* contains a single path *P* then

(2) for each combination (denoted as β) of the nodes in the path *P*

(3) generate pattern $\beta \cup \alpha$ with *support_count = minimum support count of nodes in* b;

(4) else for each *ai* in the header of *Tree{*

(5) generate pattern $\beta = ai \cup \alpha$ with *support_count = ai:support count*;

(6) construct β's conditional pattern base and then β's conditional FP tree $Tree_\beta$;

(7) if $Tree_\beta \neq 0$ then

(8) call FP growth($Tree_\beta$, β); }

9.6. ICEBERG QUERIES

Iceberg queries are used in DM and with the help of these queries an aggregate function is computed over an attribute or set of attributes above some specified threshold. Apriori algorithm is used to improve the efficiency of answering the iceberg queries. For example,

```
Select...................attributes...........
From.....................relation name
Group by.............. .
Having ...............(aggregate function>=threshold)
```

Mining multilevel association rules from transactional databases.

Till now we have discussed mining single dimensional boolean association rules from transactional databases. Now in this topic we will be discussing about Mining multilevel association rules from transactional databases.

Data mining systems provide capabilities for mining association rules at multiple levels of abstraction, with sufficient flexibility for easy traversal among different abstraction spaces. Since it is not always possible to find the strong associations among data items at low level of abstraction, therefore mining is done at multilevel. Association rules generated from mining data at multiple levels of abstraction are called multiple-level or multilevel association rules.

For efficiently mining the multilevel association rules, concept hierarchies under a support-confidence framework is used. In general, a top-down strategy is employed, where counts are accumulated for the calculation of frequent itemsets at each concept level, starting at the concept level and working downward in the hierarchy toward the more specific concept levels, until no more frequent itemsets can be found.

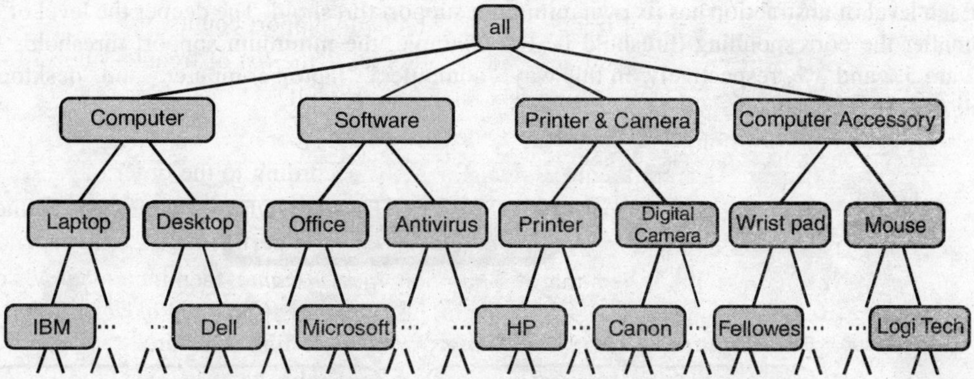

Fig. 9.1. Concept Hierarchy

9.7. DIFFERENT APPROACHES FOR MINING MULTILEVEL ASSOCIATION RULE

For each level, in concept hierarchy , any algorithm for discovering frequent itemsets may be used, such as Apriori or its variations. A number of variations to this approach:

1. Uniform Support: Using Uniform Minimum Support for all Levels

The same minimum support threshold is used when mining at each level of abstraction. For example, a minimum support threshold of 5% is used throughout. (For example, for mining from "computer" down to "laptop computer").

Advantages:

(*i*) Search procedure is simplified.

(*ii*) Avoids examining itemsets containing any item whose ancestors do not have minimum support.

Disadvantages:

(*i*) It is unlikely that items at lower levels of abstraction will occur as frequently as those at higher levels of abstraction.

(*ii*) If the minimum support threshold is set too high, it could miss some meaningful associations occurring at low abstraction levels.

(*iii*) If the threshold is set too low, it may generate many uninteresting associations occurring at high abstraction levels.

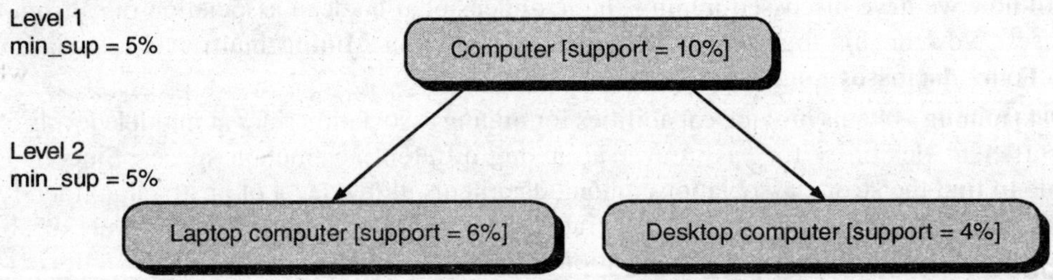

Fig. 9.2. Multilevel Mining With Uniform Support.

2. Reduced Support: Using Reduced Minimum Support at Lower Levels

Each level of abstraction has its own minimum support threshold. The deeper the level of abstraction, the smaller the corresponding threshold is. For example, the minimum support thresholds for levels 1 and 2 are 5% and 3%, respectively. In this way, "computer," "laptop computer," and "desktop computer" are all considered frequent.

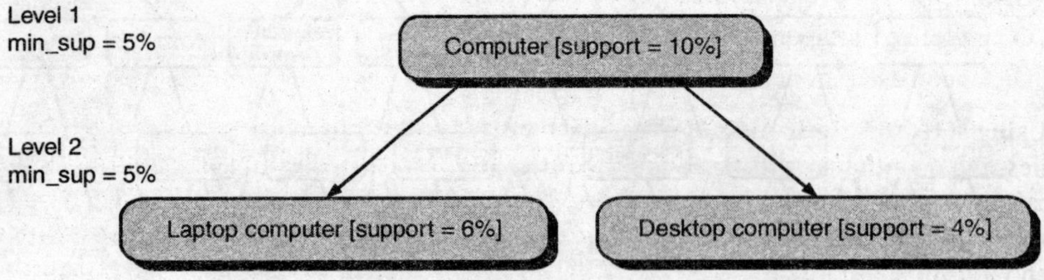

Fig. 9.3. Multilevel Mining With Reduced Support.

3. Group based support: Using item or group-based minimum support

Because users or experts often have insight as to which groups are more important than others, it is sometimes more desirable to set up user-specific, item, or group based minimal support thresholds when mining multilevel rules. For example, a user could set up the minimum support thresholds based on product price, or on items of interest, such as by setting particularly low support thresholds for laptop computers and flash drives in order to pay particular attention to the association patterns containing items in these categories.

9.8. MINING MULTI DIMENSIONAL ASSOCIATION RULES FROM RELATIONAL DATABASES AND DATA WAREHOUSE

Association rules that involve two or more dimensions or predicates can be referred to as multidimensional association rules.

1. **Single dimensional or intra dimensional association rule**: It is a rule that contains a single distinct predicate with multiple occurrences (*i.e.*, the predicate occurs more than once within the rule).

```
buys(X, "digital camera") → buys(X, "HP printer").
```

2. **Multidimensional or Inter-dimensional association rules: association rules that involve 2 or more dimensions or predicate are called Multidimensional association rules.** Here each of the predicates occur only once in the rule. Hence, we say that it has no repeated predicates.

```
age(X, "20…29")^occupation(X, "student") → buys(X, "laptop").
```

3. **Hybrid-dimensional association rules:** These rules have repeated predicates, which contain multiple occurrences of some predicates. An example of such a rule is the following, where the predicate buys is repeated.

```
age(X, "20…29")^buys(X, "laptop") ?buys(X, "HP printer")
```

Database Attributes can be Categorical or Quantitative.

1. **Categorical attributes:** They have a finite number of possible values, with no ordering among the values (e.g., occupation, brand, color). Categorical attributes are also called nominal attributes, because their values are "names of things."

2. **Quantitative attributes:** They are numeric and have an implicit ordering among values (e.g., age, income, price).

9.9. BASIC APPROACHES FOR MINING MULTIDIMENSIONAL ASSOCIATION RULES

Three basic approaches for mining multidimensional association rules are:

1. Using static discretization of quantitative attributes:
2. Quantitative association rules
3. Distance based association rules.

1. **Using static discretization of quantitative attributes:** In this approach using predefined concept hierarchies, the quantitative attributes are discretized. This process occurs prior to mining to replace the numerical values into ranges. Now by doing this each range is considered a category. But here the discretization is static and predetermined thus the name. if the task relevant data is stored in a relational table then with some modifications APRIORI algorithm is applied to find the frequent predicate sets. The transformed data can be stored in data cubes and since data cubes are multidimensional therefore these data cubes are well suited for mining of multidimensional association rules.

2. Quantitative association rules: In this the numeric attributes are dynamically discretized rather than static(as above) thus satisfying the criteria such as maximizing the confidence of the rules mined. Quantitative attributes are discretized into "bins". This is done based on how data is distributed.

3. Distance based association rules: To capture the semantic meaning of interval data, the quantitative attributes are discretized. The dynamic discretization procedure considers the distance between data points, thus the name.

Constraint Based Association Mining

A data mining process may uncover thousands of rules from a given set of data, most of which end up being unrelated or uninteresting to the users. A good heuristic is to have the users specify such intuition or expectations as constraints to confine the search space. This strategy is known as constraint-based mining. The constraints can include the following:

(*i*) **Knowledge type constraints:** These specify the type of knowledge to be mined, such as association or correlation.

(*ii*) **Data constraints:** These specify the set of task-relevant data.

(*iii*) **Dimension/level constraints:** These specify the desired dimensions (or attributes) of the data, or levels of the concept hierarchies, to be used in mining.

(*iv*) **Interestingness constraints:** These specify thresholds on statistical measures of rule interestingness, such as support, confidence, and correlation.

(*v*) **Rule constraints:** These specify the form of rules to be mined. Such constraints may be expressed as metarules (rule templates), as the maximum or minimum number of predicates that can occur in the rule antecedent or consequent, or as relationships among attributes, attribute values, and/ or aggregates.

The above constraints can be specified using a high-level declarative data mining query language and user interface.

1. Metarule-Guided Mining of Association Rules:

(*i*) Metarules allow users to specify the syntactic form of rules that they are interested in mining. The rule forms can be used as constraints to help improve the efficiency of the mining process.

(*ii*) Metarules may be based on the analyst's experience, expectations, or intuition regarding the data or may be automatically generated based on the database schema.

Example: Suppose one is interested in finding associations between customer traits and the items that customers buy - determining which pairs of customer traits promote the sale of office software.

Meta rule will be:

```
P1(X, Y)^P2(X,W) → buys(X, "office software")
```

where P1 and P2 are predicate variables that are instantiated to attributes from the given database during the mining process, X is a variable representing a customer, and Y and W take on values of the attributes assigned to P1 and P2, respectively. Typically, a user will specify a list of attributes to be considered for instantiation with P1 and P2.

(*iii*) The data mining system can then search for rules that match the given metarule. For instance, this rule matches or complies with earlier metarule.

```
age(X, "30…39")^income(X, "41K…60K") →buys(X, "office software")
```

Metarules to guide mining process:

(*i*) A metarule is a rule template of the form

$$P_1 \wedge P_2 \wedge \ldots \wedge P_l \rightarrow Q1 \wedge Q2 \wedge \ldots . \wedge Qr$$

where Pi (i = 1, ..., l) and Qj (j = 1, ..., r) are either instantiated predicates or predicate variables.

Let the number of predicates in the metarule be p = l + r. To find interdimensional association rules satisfying the template, find all frequent p-predicate sets, Lp have the support or count of the l-predicate subsets of Lp in order to compute the confidence of rules derived from Lp.

1. Constraint Pushing: Mining Guided by Rule Constraints

Rule constraints specify expected set/subset relationships of the variables in the mined rules, constant initiation of variables, and aggregate functions. These rule constraints may be used with metarule-guided mining. For example, multidimensional database with the following interrelated relations:

Dimension Tables:

```
sales(customer name, item name, TID)
lives in(customer name, region, city)
item(item name, group, price)
transaction(TID, day, month, year)
```

Fact table sales :

```
customer name, item name, and TID (transaction id).
```

Mining query is :

"Find the sales of which cheap items (where the sum of the prices is less than $100) may promote the sales of which expensive items (where the minimum price is $500) of the same group for Chicago customers in 2004."

This can be expressed in DMQL:

```
(1) mine associations as
(2) lives in(C, , "Chicago") ^ sales +(C,?{I},{S}) sales+(C, ,?{I},{S})
(3) from sales
(4) where S.year = 2004 and T.year = 2004 and I.group = J.group
(5) group by C, I.group
(6) having sum(I.price) < 100 and min(J.price) > 500
(7) with support threshold = 1%
(8) with confidence threshold = 50%
```

Line 1 is a knowledge type constraint, where association patterns are to be discovered. Line 2 specifies a metarule. Dimension/level constraints and interestingness constraints can be applied after mining to filter out discovered rules, although it is generally more efficient and less expensive to use them during mining, to help prune the search space.

2. Classification of Rule Constraints

Rule constraints can be classified into the following five categories with respect to frequent itemset mining: (1) antimonotonic, (2) monotonic, (3) succinct, (4) convertible, and (5) inconvertible.

Antimonotonic: If an itemset does not satisfy this rule constraint, none of its supersets can satisfy the constraint. If a rule constraint obeys this property, it is antimonotonic.

```
"avg(I.price) <= 100" is not antimonotonic.
```

For a given itemset that does not satisfy this constraint, a superset created by adding some (cheap) items may result in satisfying the constraint. Pruning by antimonotonic constraints can be applied at each iteration of Apriori-style algorithms to help improve the efficiency of the overall mining process while guaranteeing completeness of the data mining task.

Monotonic: If an itemset satisfies this rule constraint, so do all of its supersets. If a rule constraint obeys this property, it is monotonic. Example:

```
"sum(I.price) > 100,"
```

If an itemset I satisfies the constraint, that is, the sum of the prices in the set is no less than 100, further addition of more items to I will increase cost and will always satisfy the constraint.

Succinct: All and only those set guaranteed to satisfy the rule can be enumerated. The itemsets can be generated that satify the rule even before support counting begins. Once such subset is generated, iterative testing for the constraint can be effectively avoided. For example, the constraint

```
"min(J.price)>500" , "Max(S)<120"
```

avg(S)>v, avg(S)<v are not succinct.

Convertible: If the items in the itemset are arranged in a particular order, the constraint may become monotonic or antimonotonic with regard to the frequent itemset mining process. For example, the constraint

```
"avg(I.price) <= 100"
```

is neither antimonotonic nor monotonic. However, if items in a transaction are added to an itemset in price-ascending order, the constraint becomes antimonotonic, because if an itemset I violates the constraint (*i.e.*,with an average price greater than $100), then further addition of more expensive items into the itemset will never make it satisfy the constraint.

Inconvertible: Constraints which are not convertible.

For example,

```
"sum(S)<v," , "sum(S)<V,"
```

where element in S could be of any real value, is not convertible.

Summary

- Association rule mining has been extensively studied in the data mining community, To discover the relationships among huge amounts of data.
- Rule strength measures are i) Support $sup = \Pr(X \cup Y)$.
  Confidence $conf = \Pr(Y \mid X)$
- Association rules can be classified in various ways, discussed below
  1. Based on the types of values handled in the rule
  2. Based on the dimensions of data involved in the rule
  3. Based on the levels of abstraction involved in the rule
  4. Based on various extension to association mining
- Two steps in apriori algorithm are:
  (*i*) Find all itemsets that have minimum support (*frequent itemsets*, also called large itemsets).
  (*ii*) Use frequent itemsets to generate rules.
- To make apriori algorithm more efficient hasing, transaction reduction can be used.
- **Frequent Pattern Mining:** Frequent pattern mining searches for recurring relationships in a given data set. Frequent itemset mining leads to the discovery of associations and correlations among items in large transactional or relational data sets. The discovery of interesting correlation relationships among huge amounts of business transaction records can help in many business

decision-making processes, such as catalog design, cross-marketing, and customer shopping behavior analysis.

Example: Market basket analysis is just one form of frequent pattern mining.

- Techniques for multidimensional association rule can be categorized according to how quantitative attributes are used. Three techniques for mining multidimensional association rules are:
  1. Using static discretization of quantitative attributes:
  2. Quantitative association rules
  3. Distance based association rules.

Review Questions

1. Explain the concept of association rule mining with the help of example.
2. How the rules of strength are measured, explain.
3. Discuss the classification of association rules.
4. Explain Apriori Algorithm with its advancements.
5. Explain mining association rules form multidimensional data.

❑❑❑

10

Chapter

Clustering

INSIDE THIS CHAPTER

10.1. Introduction **10.2.** Applications of Clustering **10.3.** Requirements of Clustering in Data Mining **10.4.** Categories of Clustering Methods **10.5.** Outlier Analysis

In this chapter we will learn meaning of clustering, various clustering methods like Partitioning methods, Hierarchical methods, Density based methods, and Model based clustering etc.

10.1. INTRODUCTION

Cluster analysis or **clustering** is the task of assigning a set of objects into groups (called **clusters**) so that the objects in the same cluster are more similar (in some sense or another) to each other than to those in other clusters.

A **cluster** is a group of relatively homogeneous cases or observations. Objects in a cluster are similar to each other. They are also dissimilar to objects outside the cluster, particularly objects in other clusters.

For example, a person might wish to predict how an animal would respond to an invitation to go for a walk. He or she could be given information about the size and weight of the animal, top speed, average number of hours spent sleeping per day, and so forth and then combine that information into a prediction of behavior. Alternatively, the person could be told that an animal is either a lion or a elephant. The latter information allows a much broader range of behaviors to be predicted. The trick in cluster analysis is to collect information and combine it in ways that allow classification into useful groups, such as lion or elephant.

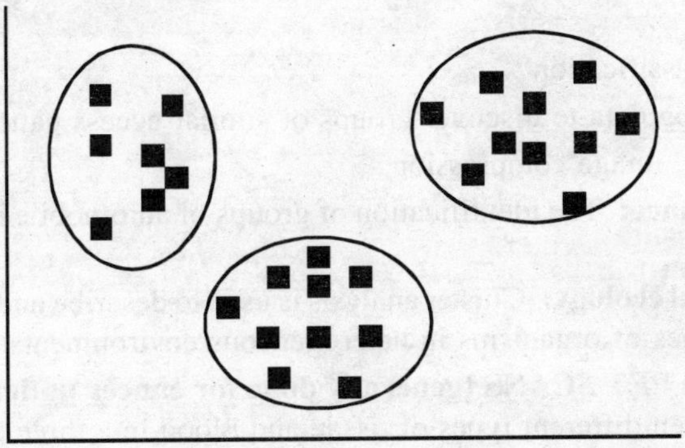

Fig. 10.1.

Many definitions have been proposed: The process of grouping a set of physical or abstract objects into classes of similar objects is called clustering.

1. A cluster is a collection of data objects that are similar to one another within the same cluster and are dissimilar to the objects in other clusters. A cluster of data objects can be treated collectively as one group and so may be considered as a form of data compression. First the set is partitioned into groups based on data similarity (e.g., using clustering), and then labels are assigned to the relatively small number of groups.

2. the distance between the points in a cluster is less than the distance between a point in the cluster and any point outside it.

Clustering is also called **unsupervised learning**. Clustering is called unsupervised learning because unlike classification, clustering and unsupervised learning do not rely on predefined classes and class-labeled training examples. That is why, clustering is a form of learning by observation, rather than learning by examples. Clustering is also called **data segmentation** in some applications because clustering partitions large data sets into groups according to their similarity

Supervised Classification = Classification: We know the class labels and the number of classes

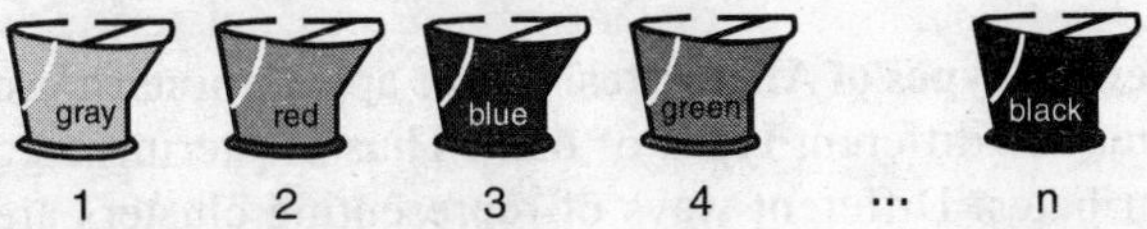

Unsupervised Classification = Clustering: We do not know the class labels and may not know the number of classes.

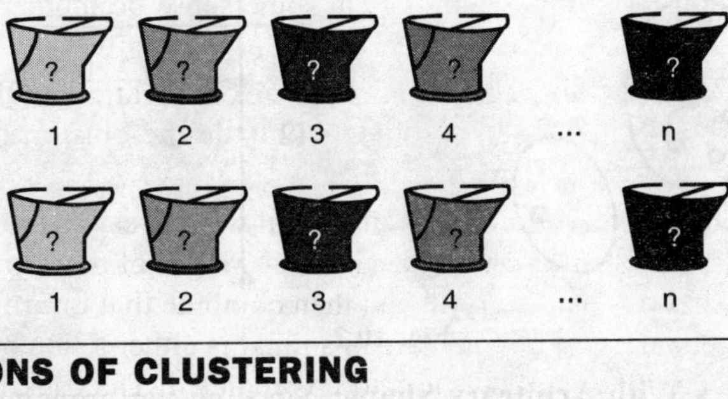

10.2. APPLICATIONS OF CLUSTERING

Clustering has wide application in

1. Spatial data analysis:

(*i*) create thermatic maps in GIS by clustering feature spaces

(*ii*) detect spatial clusters and explain them in spatial data mining.

2. WWW

(*i*) Document classification

(*ii*) Cluster Weblog data to discover groups of similar access patterns.

3. Pattern recognition, image compression.

4. Automobile insurance: The identification of groups of automobile insurance policy holders with a high average claim cost.

5. In plant and animal ecology: Cluster analysis is used to describe and to make spatial and temporal comparisons of communities of organisms in heterogeneous environments.

6. In medicine like in PET SCANS (generally done for cancer patients): Cluster analysis can be used to differentiate between different types of tissue and blood in a three dimensional image.

7. Market research: Can help marketers discover distinct groups in their customer bases and characterize customer groups based on purchasing patterns.

8. Computer science: Clustering is useful in software evolution as it helps to reduce legacy properties in code by reforming functionality that has become dispersed. It is a form of restructuring and hence is a way of directly preventative maintenance.

9. Image segmentation: Clustering can be used to divide a digital image into distinct regions for border detection or object recognition.

10. Crime analysis: Cluster analysis can be used to identify areas where there are greater incidences of particular types of crime. By identifying these distinct areas or "hot spots" where a similar crime has happened over a period of time, it is possible to manage law enforcement resources more effectively.

11. Educational data mining: Cluster analysis is for example used to identify groups of schools or students with similar properties.

10.3. REQUIREMENTS OF CLUSTERING IN DATA MINING

1. Scalability: Since the databases not always of small sizes, therefore clustering on a sample of large data set may lead to biased results, therefore to deal with this problem highly scalable algorithms are required.

2. Dealing with Different Types of Attributes: Since applications may contain numeric data, binary data, ordinal data or mixture of different types of data. Thus clustering algorithm must be able to deal with different types of attributes. Different ways of representing clusters are Division with boundaries, Sphere, Probabilistic, Dendrograms etc.

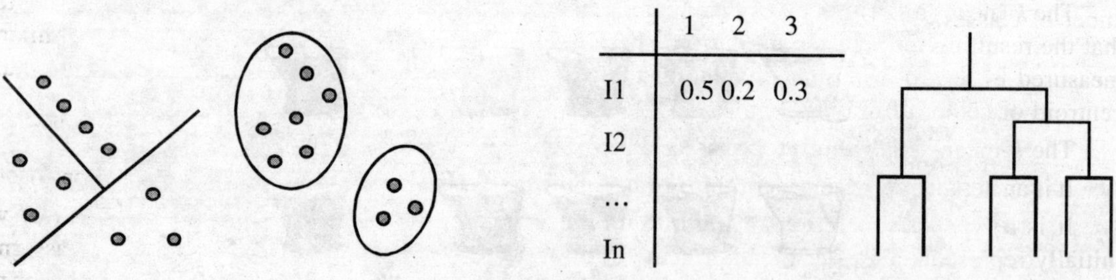

Fig. 10.2.

3. Discovery of Clusters With Arbitrary Shape: Since cluster(grouping of similar objects) can be of any shape, thus it is important to develop the algorithms that can detect clusters of any shape.

4. Minimal Requirements for Domain Knowledge to Determine Input Parameters: Many clustering algorithms require the users to input certain parameters, and the clustering results can be sensitive to input parameters. The algorithms must deal with this kind of problem.

5. Ability to Deal With Noisy Data and Outlier: The algorithms must be able to deal with noisy, missing or erroneous data since some of the clustering algorithms are sensitive to such data and gives bad results.

6. Insensitivity to the Order of Input Records: Algorithms must not be sensitive to the order of input data.

7. High Dimensionality: The clustering algorithms must be able to deal with low as well as high dimensional data.

8. Interpretability and Usability: Clustering is not used as stand–alone rather it may need to be tied up with specific applications. Thus clustering must has the property of interpretability and usable.

10.4. CATEGORIES OF CLUSTERING METHODS

Clustering methods have been used in data mining, statistics, machine learning etc. various methods have designed and proposed. Five categories of clustering methods are :

1. Partitioning algorithms: Construct various partitions and then evaluate them by some criterion.

2. Hierarchy algorithms: Create a hierarchical decomposition of the set of data (or objects) using some criterion. Thre is an agglomerative approach and a divisive approach.

3. Density-base: Based on connectivity and density functions.

4. Grid-based: Based on a multiple-level granularity structure.

5. Model-based: A model is hypothesized for each of the clusters and the idea is to find the best fit of that model to each other.

10.4.1. Partitioning Algorithms

Given a database of n objects or data tuples, a partitioning method constructs *k* partitions of the data, where each partition represents a cluster and *k* <= n. That is, data is classified into *k* groups, which together satisfy the following requirements: (1) each group must contain at least one object, and (2) each object must belong to exactly one group.

Given *k*, the number of partitions to construct, a partitioning method creates an initial partitioning. It then uses an iterative relocation technique that attempts to improve the partitioning by moving objects from one group to another. Examples (1) the *k*-means algorithm and (2) the *k*-medoids algorithm

1. Classical Portioning Method-*k*-means Method

The *k*-means algorithm takes the input parameter, *k*, and partitions a set of *n* objects into *k* clusters so that the resulting intracluster similarity is high but the intercluster similarity is low. Cluster similarity is measured in regard to the mean value of the objects in a cluster, which can be viewed as the cluster's centroid or center of gravity.

The *k*-means algorithm proceeds as follows:

It is an iterative clustering algorithm *i.e.* the following process iterates until the criterion function converges.

It is a type of squared error algorithm.First, it randomly selects *k* of the objects, each of which initially represents a cluster mean or center. For each of the remaining objects, an object is assigned to the cluster to which it is the most similar, based on the distance between the object and the cluster mean. It then computes the new mean for each cluster.

Typically, the square-error criterion is used, defined as

$$E = \sum_{i=1}^{k} \sum_{p \in Ci} | p - mi |^2$$

where E is the sum of the square error for all objects in the data set; p is the point in space representing a given object; and m_i is the mean of cluster C_i (both p and m_i are multidimensional) and mean = total sum/number of items.

Algorithm:

Input : k is the number of clusters, and D is a data set containing n objects.

Output: A set of k clusters.

Method:

(1) arbitrarily choose k objects from D as the initial cluster centers;

(2) repeat

(3) (re)assign each object to the cluster to which the object is the most similar, based on the mean value of the objects in the cluster;

(4) update the cluster means, i.e., calculate the mean value of the objects for each cluster;

(5) until no change;

Advantages/ Strengths

(*i*) This algorithm works well when the clusters are compact clouds that are rather well separated from one another.

(*ii*) Clusters are relatively scalable and efficient in processing. The complexity of the algorithm is $O(nkt)$, where n is the total number of data points, k is the number of clusters, and t is the number of iterations. Normally k, $t << n$.

Limitations/Weakness

(*i*) K-means algorithm can be applied only when the mean of a cluster is defined.

(*ii*) The necessity for users to specify k in advance can be seen as a disadvantage.

(*iii*) This algorithm is not suitable for discovering clusters with nonconvex shapes or clusters of very different size.

(*iv*) K-means algorithm is Sensitive to noise and outlier data points because a small number of such data can substantially influence the mean value.

Example 10.1.

Input: 1 dimensional data and *k*=2.

Data: 1, 2, 5, 6,7

Solution. K-means:

Randomly select 5 and 6 as initial centroids;

=> Two clusters {1,2,5} and {6,7}; mean C1=(1+2+5)/3=2.6, meanC2=(6+7)/2=6.5

=> again 2 clusters {1,2}, {5,6,7}; meanC1=1.5, meanC2=6

=> no change.

Example 10.2. Input = {2, 10, 5, 4, 25, 15, 20, 40} and k = 2.

Solution.

Step 1: Break the set into 2 clusters. Randomly assign data into two clusters and calculate the mean value:

K1: {2, 10, 5, 4} Mean = 5.25

K2: {25, 15, 20, 40} Mean = 25

Step 2: Reassign the values to the numbers which are close to the mean of K1(5.25) to K1, mean of K2(25) to K2.

K1:{2,5,4} Mean = 3.66

K2:{10,25,15,20,40} Mean = 22

Step 3: Reassign

K1:{2,5,4} Mean = 3.66

K2:{10,25,15,20,40} Mean = 22

Both the clusters are the same as in Step 2.

So, the final answer is

K1: {2,5,4}

K2:{10,25,15,20,40}

Example 10.3. Use K-means algorithm to create 3 clusters for a set of values {2, 3, 6, 8, 9, 12, 15, 18, 22}

Solution.

Step 1: Break the set into 3 clusters. Randomly assign data to three clusters and calculate the mean value:

K1: 2,8,15 – mean 8

K2: 3,9,18 – mean 10

K3: 6,12,22 – mean 13.3

Step 2: Reassign the values to the numbers which are close to the mean of K1(8) to K1, mean of K2(10) to K2, mean of K3(13.3) to K3.

K1: 2,3,6,8,9 – mean 5.6

K2: - mean 0

K3: - 12,15,18,22 – mean 16.75

Step 3: Reassign

K1: 6,8,9 – mean 7.6

K2: 2,3 – mean 2.5

K3: 12,15,18,22 – mean 16.75

Step 4:Reassign

K1: 6,8,9 – mean 7.6

K2: 2,3 – mean 2.5

K3: 12,15,18,22 – mean 16.75

Last two groups are same. 3 clusters are:

Cluster 1= {6,8,9}, Cluster 2 ={2,3} , Cluster 3={12,15,18,22}

2. K-Medoids Method

Due to limitation of *k*-means method that it is sensitive to outlier (explained above) another method called *k*-medoids is used. The basic principal of this algorithm is

(*i*) Instead of taking the mean value of the objects in a cluster as a reference point, actual objects can be picked to represent the clusters, using one representative object per cluster.

(*ii*) Each remaining object is clustered with the representative object to which it is the most similar.

(*iii*) The partitioning method is then performed based on the principle of minimizing the sum of the dissimilarities between each object (cost function) and its corresponding reference point.

(*iv*) In general, the algorithm iterates until, eventually, each representative object is actually the medoid, or most centrally located object, of its cluster.

k-medoids algorithm

Input: the number of clusters and a database containing *n* objects.

Output: A set of *k* clusters that minimizes the sum of the dissimilarities of all the objects to their nearest medoid.

Steps:

1. arbitrarily choose *k* objects from D as the initial cluster centers;
   Repeat
2. assign each remaining object to the cluster with the nearest medoid
3. randomly select a nonmedoid object O_{rand}
4. Compute the total cost S by swaping medoid object with non medoid object.
5. if S < 0, then swap medoid object with non medoid object.
   To form the new set of *k* medoids.
6. until no change;

PAM(Partitioning Around Medoids) was one of the first *k*-medoids algorithms introduced. It attempts to determine *k* partitions for *n* objects.

Use real object to represent the cluster.

(*i*) Select *k* representative objects arbitrarily

(*ii*) For each pair of non-selected object *h* and selected object *i*,
Calculate the total swapping cost TC_{ih}.
If $TC_{ih} < 0$, *i* is replaced by *h*.

(*iii*) Then assign each non-selected object to the most similar representative object.

(*iv*) Repeat steps 2-3 until there is no change.

Comments on PAM

(*i*) Pam is more robust than *k*-means in the presence of noise and outliers because a medoid is less influenced by outliers or other extreme values than a mean and

(*ii*) Pam works well for small data sets but does not scale well for large data sets. The complexity of PAM is $O(k(n-k)^2)$ for each change where *n* is total number of data, *k* is number of clusters.

10.4.2. Hierarchical Methods

Hierarchical clustering is a method of cluster analysis which seeks to build a hierarchy of clusters. Strategies for hierarchical clustering generally fall into two types:

(*i*) **Agglomerative:** This is a "bottom up" approach: each observation starts in its own cluster, and pairs of clusters are merged as one moves up the hierarchy.

(*ii*) **Divisive:** This is a "top down" approach: all observations start in one cluster, and splits are performed recursively as one moves down the hierarchy.

Hierarchical methods use distance matrix for clustering. This method does not require the number of clusters *k* as an input, but needs a termination condition.

The results of hierarchical clustering are usually presented in a dendrogram. Dendogram is a tree diagram frequently used to illustrate the arrangement of the clusters produced by hierarchical clustering. Dendrograms are often used in computational biology to illustrate the clustering of genes or samples.

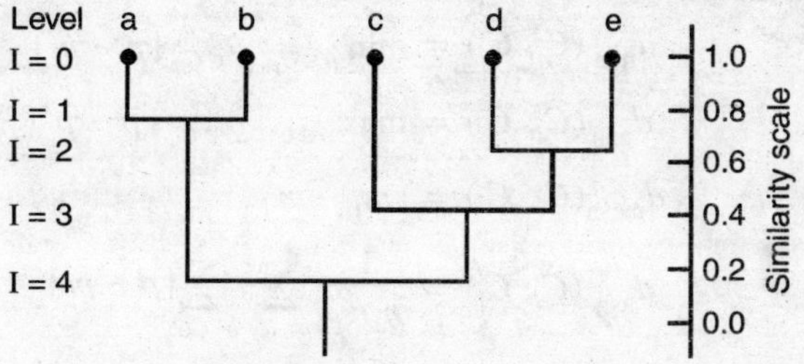

Fig. 10.3. Dendrogram Representation for Hierarchical Clustering of Data Objects {a, b, c, d, e}.

Some commonly used metrics for hierarchical clustering are:

Names	**Formula**
Euclidean distance	$\|\|a-b\|\|_2 = \sqrt{\sum_i (a_i - b_i)^2}$
Squared Euclidean distance	$\|\|a-b\|\|_2^2 = \sum_i (a_i - b_i)^2$
Manhattan distance	$\|\|a-b\|\|_1 = \sum_i \|a_i - b_i\|^2$
Maximum distance	$\|\|a-b\|\|_\infty = \max_i \|a_i - b_i\|$
Mahalanobis distance	$\sqrt{(a-b)^{T} S^{-1} (a-b)}$ where S is the covariance matrix
Cosine similarity	$\frac{a \cdot b}{\|\|a\|\|\,\|\|b\|\|}$

For text or other non-numeric data, metrics such as the Hamming distance or Levenshtein distance are often used.

Example 10.5. Let $x1 = (1, 2)$ and $x2 = (3, 5)$ represent two objects as in Figure. Find the Euclidean distance and Manhattan distance.

Solution. The Euclidean distance between the two is

$$\sqrt{(2^2 + 3^2)} = 3.61.$$

The Manhattan distance between the two is 2+3 = 5.

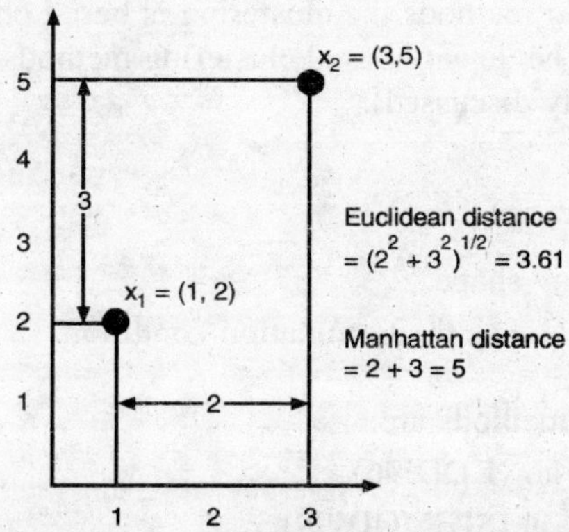

Fig. 10.4. Euclidean and Manhattan distances between two objects

In either agglomerative or divisive hierarchical clustering, the user can specify the desired number of clusters as a termination condition. Four widely used measures for distance between clusters are as follows, where | p – p′ | is the distance between two objects or points, p and p′; mi is the mean for cluster,Ci; and ni is the number of objects in Ci.

Minimum distance : $d_{min}(C_i, C_j) = \min_{p \in C_i, p' \in C_j} |p - p'|$

Maximum distance : $d_{max}(C_i, C_j) = \max_{p \in C_i, p' \in C_j} |p - p'|$

Mean distance : $d_{mean}(C_i, C_j) = |m_i - m_j|$

Average distance : $d_{avg}(C_i, C_j) = \frac{1}{n_i n_j} \sum_{p \in C_i} \sum_{p' \in C_j} |p - p'|$

Nearest–neighbor clustering: When an algorithm uses the minimum distance, dmin(Ci, Cj), it is called a nearest-neighbor clustering algorithm. Moreover, if the clustering process is terminated when the distance between nearest clusters exceeds an arbitrary threshold, it is called a single-linkage algorithm. An agglomerative hierarchical clustering algorithm that uses the minimum distance measure is also called a minimal spanning tree algorithm.

farthest-neighbor clustering algorithm: When an algorithm uses the maximum distance, dmax(Ci, Cj), to measure the distance between clusters, it is sometimes called a farthest-neighbor clustering algorithm. If the clustering process is terminated when the maximum distance between nearest clusters exceeds an arbitrary threshold, it is called a complete-linkage algorithm

The above minimum and maximum measures represent two extremes in measuring the distance between clusters. They tend to be overly sensitive to outliers or noisy data. The use of mean or average distance is a compromise between the minimum and maximum distances and overcomes the outlier sensitivity problem. Whereas the mean distance is the simplest to compute, the average distance is advantageous in that it can handle categoric as well as numeric data. The computation of the mean vector for categoric data can be difficult or impossible to define.

Advantages: The advantages of hierarchical clustering algorithms are their simplicity.

Limitations of Hierarchical Clustering: Still there are some limitations like

1. Difficulties regarding the selection of merge or split points.
2. It will neither undo what was done previously nor perform object swapping between clusters. Thus merge or split decisions, if not well chosen at some step, may lead to low-quality clusters.
3. The method does not scale well. that is scalability is a problem.

10.4.3. Density Based Clustering Methods

In density based clustering methods the clustering is based on density(local cluster criterion) such as density connected points. The density based clustering methods are used to discover the clusters with arbitrary shape (shapes already discussed).

Features:

1. Handle noise
2. Discover cluster of any shape.
3. Need density parameters as the termination condition.

Density based clustering methods are:

1. DBSCAN: Ester, et. al. (KDD'96)
2. OPTICS: Ankerst, et. al (SIGMOD'99).

3. DENCLUE: Hinneburg and D. Keim (KDD'98).
4. CLIQUE: Agrawal, et. al. (SIGMOD'98).

DBSCAN: density based clustering method based on connected regions with sufficiently high density.

DBSCAN's definition of a cluster is based on the notion of *density reachability*. Basically, a point *q* is directly density-reachable from a point *p* if it is not farther away than a given distance *e* (*i.e.*, is a part of its ε-neighbourhood) and if p is surrounded by sufficiently many points such that one may consider p and q to the part of a cluster. q is called *density-reachable* (note the distinction from "directly density-reachable") from p if there is a sequence $p_1, ..., p_n$ of points with $p_1 = p$ and pn = q where each p_{i+1} is directly density-reachable from p_i.

Note that the relation of density-reachable is not symmetric. q might lie on the edge of a cluster, having insufficiently many neighbors to count as dense itself. This would halt the process of finding a path that stops with the first non-dense point. By contrast, starting the process with q *would* lead to p (though the process would halt there, p being the first non-dense point). Due to this asymmetry, the notion of *density-connected* is introduced: two points p and q are density-connected if there is a point o such that both p and q are density-reachable from o. Density connectedness is symmetric.

A cluster, which is a subset of the points of the database, satisfies two properties:

1. All points within the cluster are mutually density-connected.
2. If a point is density-connected to any point of the cluster, it is part of the cluster as well.

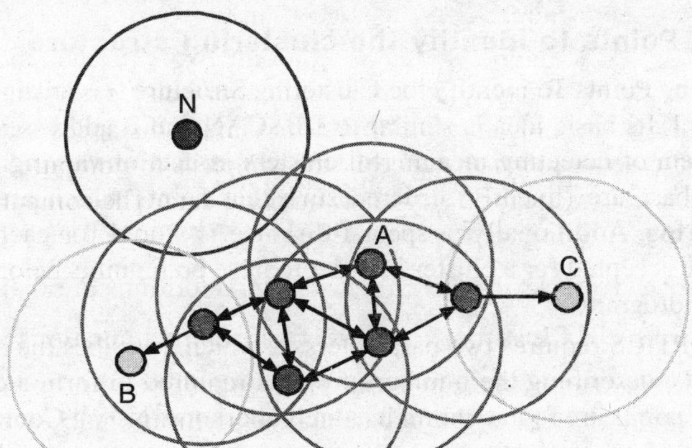

Fig. 10.5.

Algorithm

DBSCAN requires two parameters: ε (eps) and the minimum number of points required to form a cluster (minPts). It starts with an arbitrary starting point that has not been visited. This point's ε-neighborhood is retrieved, and if it contains sufficiently many points, a cluster is started. Otherwise, the point is labeled as noise. Note that this point might later be found in a sufficiently sized ε-environment of a different point and hence be made part of a cluster.

If a point is found to be a dense part of a cluster, its ε-neighborhood is also part of that cluster. Hence, all points that are found within the ε-neighborhood are added, as is their own ε-neighborhood when they are also dense. This process continues until the density-connected cluster is completely found. Then, a new unvisited point is retrieved and processed, leading to the discovery of a further cluster or noise.

Advantages

1. DBSCAN does not require one to specify the number of clusters in the data a priori, as opposed to k-means.
2. DBSCAN can find arbitrarily shaped clusters. It can even find a cluster completely surrounded by (but not connected to) a different cluster. Due to the MinPts parameter, the so-called single-link effect (different clusters being connected by a thin line of points) is reduced.
3. DBSCAN has a notion of noise.
4. DBSCAN requires just two parameters and is mostly insensitive to the ordering of the points in the database. (However, points sitting on the edge of two different clusters might swap cluster membership if the ordering of the points is changed, and the cluster assignment is unique only up to isomorphism.)

Disadvantages:

The quality of DBSCAN depends on the distance measure used in the function regionQuery(P, ε). The most common distance metric used is Euclidean distance. Especially for high-dimensional data, this metric can be rendered almost useless due to the so-called "Curse of dimensionality", making it difficult to find an appropriate value for ε .,. This effect, however, is also present in any other algorithm based on Euclidean distance. DBSCAN cannot cluster data sets well with large differences in densities, since the minPts-ε combination cannot then be chosen appropriately for all clusters.

OPTICS: Ordering Points to identify the clustering structure

OPTICS ("Ordering Points To Identify the Clustering Structure") is an algorithm for finding density-based clusters in spatial .Its basic idea is similar to DBSCAN, but it addresses one of DBSCAN's major weaknesses: the problem of detecting meaningful clusters in data of varying density. In order to do so, the points of the database are (linearly) ordered such that points which are spatially closest become neighbors in the ordering. Additionally, a special distance is stored for each point that represents the density that needs to be accepted for a cluster in order to have both points belong to the same cluster. This is represented as a dendrogram.

Like DBSCAN, OPTICS requires two parameters: ε, which describes the maximum distance (radius) to consider, and MinPts, describing the number of points required to form a cluster. A point *p* is a *core point* if at least MinPts points are found within its ε-neighborhood Nε(p). Contrary to DBSCAN, **OPTICS** also considers points that are part of a more densely packed cluster, so each point is assigned a *core distance* that basically describes the distance to its MinPtsth point:

$$core-distance_{\varepsilon,\, MinPts}(p) = \begin{cases} \text{Undefined} & \text{if } |N_{\varepsilon}(p)| < MinPts \\ \text{distance to the } MinPts\text{ - th point} & \text{otherwise} \end{cases}$$

The *reachability-distance* of a point *p* from another point *o* is the distance between *p* and *o*, or the core distance of *o*:

$$reachability\text{ - }distance_{\varepsilon,\, MinPts}(p, o) = \begin{cases} \text{UNDEFINED} \\ \max(\text{core - distance}_{\varepsilon,\, MinPts}(o), \text{distance} \end{cases}$$

Basically, if *o* and *p* are nearest neighbors, this is the $\varepsilon' < \varepsilon$, we need to assume in order to have o and p belong to the same cluster.

Both the core-distance and the reachability-distance are undefined if no sufficiently dense cluster (w.r.t., ε) is available. Given a sufficiently large ε, this will never happen, but then every ε-neighbourhood query will return the entire database, resulting in $O(n^2)$ runtime. Hence, the ε parameter is required to cutoff the density of clusters.

10.4.4. Model Based Clustering Methods

Model based clustering methods use certain models for clusters and attempt to optimize the fitness of data and the model. One of the approach followed by model based clustering method is neural network approach.

Neural Network Approach: The best known neural network approach to clustering is the SOM (*self-organizing feature map*) method, proposed by Kohonen in 1981. It can be viewed as a nonlinear projection from an m-dimensional input space onto a lower-order (typically 2-dimensional) regular lattice of cells. Such a mapping is used to identify clusters of elements that are similar (in a Euclidean sense) in the original space.

Other approach is competitive learning or winner -take-all learning. **Winner-take-all** is a computational principle applied in computational models of neural networks by which neurons in a layer compete with each others for activation. In the classical form, only the neuron with the highest activation stays active while all other neurons shut down.

10.5. OUTLIER ANALYSIS

The general meaning of outlier is " the person or thing excluded from the group that is outsider".

Outlier Analysis : A database may contain data objects that do not comply with the general behavior or model of the data. These data objects are outliers.

Example: Use in finding Fraudulent usage of credit cards. Outlier Analysis may uncover Fraudulent usage of credit cards by detecting purchases of extremely large amounts for a given account number in comparison to regular charges incurred by the same account. Outlier values may also be detected with respect to the location and type of purchase or the purchase frequency.

Outlier detection and analysis is data mining task referred as OUTLIER MINING OR OUTLIER DETECTION.

Two basic types of procedures for detecting outliers :

1. Block procedures: In this case, either all of the suspect objects are treated as outliers or all of them are accepted as consistent.

2. Sequectial Procedure: This procedure is more effective than block procedure. In this case, the object that is least " likely" to be outlier is tested first, if it is found to be an outlier, then all of the extreme values are also considered outliers. Otherwise, the next most extreme object is tested and so on.

Some efficient algorithms for mining distance-based outliers are as follows:

1. Index-based algorithm
2. Nested-loop algorithm;
3. Cell-based algorithm.
4. Density-Based local outlier detection
5. Deviation based outlier detection with sequential exception technique

Problems and Challenges in Clustering

Considerable progress has been made in scalable clustering methods:

Partitioning: k-means, k-medoids, CLARANS

Hierarchical: BIRACH, CURE

Density-based: DBSCAN, CLIQUE, OPTICS

Grid-based: STING, WaveCluster.

Model-based: Autoclass, Denclue, Cobweb.

Current clustering techniques do not address all the requirements adequately (and concurrently)

Large number of dimensions and large number of data items. Strict clusters vs overlapping clusters.

Summary

- Clustering is a process of partitioning a set of data (or objects) in a set of meaningful sub-classes, called **clusters.**
- Unsupervised learning is used in clustering.
- Good clustering: A good clustering method will produce high quality clusters in which:

  the **intra-class** (that is, intra-cluster) similarity is high.

  the **inter-class** similarity is low.

  The quality of a clustering result also depends on both the similarly measure used by the method and its implementation.

  The quality of a clustering method is also measured by its ability to discover some or all of the **hidden** patterns.
- Clustering analysis found its applications in market research, biological studies, WWW, pattern recognition, image processing etc.
- Clustering methods can be classified into 5 categories as:

  1. Partitioning algorithms: Construct various partitions and then evaluate them by some criterion.

  2. Hierarchy algorithms: Create a hierarchical decomposition of the set of data (or objects) using some criterion. Thre is an agglomerative approach and a divisive approach.

  3. Density-base: Based on connectivity and density functions.

  4. Grid-based: Based on a multiple-level granularity structure.

  5. Model-based: A model is hypothesized for each of the clusters and the idea is to find the best fit of that model to each other.
- **Partitioning method:** Construct a partition of a database *D* of *n* objects into a set of k clusters.

  Given a *k,* find a partition of *k clusters* that optimizes the chosen partitioning criterion.

  Global optimal: exhaustively enumerate all partitions.

  Heuristic methods: k-means and k-medoids algorithms.

  k-means (MacQueen'67): Each cluster is represented by the center of the cluster.

  k-medoids or PAM (Partition around medoids) (Kaufman and Rousseaw)

  Each cluster is represented by one of the objects in the cluster.
- **Hierarchical clustering method: Two types of hierarchical clustering Algorithms**

  1. Agglomerative (bottom-up) merge clusters iterative. Start by placing each object in its own cluster. Merge these atomic clusters into larger and larger clusters. Until all objects are in a single cluster. Most hierarchical methods belong to this category. They differ only in their definition of between-cluster similarity.

  Divisive (top-down) split a cluster iteratively.

  It does the reverse by starting with all objects in one cluster and subdividing them into smaller pieces.

  Divisive methods are not generally available, and rarely have been applied.

  1. Model based clustering methods use certain models for clusters and attempt to optimize the fitness of data and the model. One of the approach followed by model based clustering method is neural network approach(self organizing feature maps, competitive learning).

2. Outlier detection and analysis is another main area where DM is used. It is useful for fraud detection, medical analysis etc.

Review Questions

1. What do u mean by clustering, explain with example.
2. Write the applications of clustering.
3. What is the difference between Clustering and classification, explain.
4. Clustering use unsupervised learning while classification use supervised learning, justify the statement.
5. Explain k-means and k-medoids algorithm with an example.
6. Explain outlier analysis with its application.
7. Discuss the neural network approach for model based clustering.
8. Explain hierarchical based clustering. What is the different between its two main types.
9. Explain density based algorithms for clustering.
10. Discuss the problems and the challenges in Clustering algorithms.

□□□

11

Chapter

Mining Complex Types of Data

INSIDE THIS CHAPTER

11.1. Introduction **11.2.** Spatial Database Mining **11.3.** Web Mining **11.4.** Text Mining **11.5.** Multimedia Databases Mining **11.6.** Mining Time Series and Sequence Data **11.7.** Trends in Data Mining **11.8.** Mining Time Series Data **11.9.** Complex Data Objects Mining **11.10.** Sequence Pattern Mining

This chapter deals with an important task of data mining that is to mine the complex types of data, including spatial databases, multimedia databases, time series analysis, sequence data modeling, text databases and web mining.

11.1. INTRODUCTION

The overall goal of the data mining process is to extract knowledge from an existing data set and transform it into a human-understandable structure for further use. Besides the raw analysis step, it involves database and data management aspects, data preprocessing, model and inference considerations, interestingness metrics, complexity considerations, post-processing of found structures, visualization, and online updating. In this chapter, we will study the data mining methods that handle complex data objects, spatial, multimedia, text, and Web data. These kinds of data are commonly encountered in many social, economic, scientific, engineering, and governmental applications, and pose new challenges in data mining.

11.2. SPATIAL DATABASE MINING

Spatial data mining is the process of discovering interesting, useful, non-trivial patterns from large spatial datasets.

A **spatial database** is a database that is optimized to store and query data that is related to objects in **space**, including points, lines and polygons. While typical databases can understand various numeric and character types of data, additional functionality needs to be added for databases to process spatial data types. These are typically called *geometry* or *feature*.

A spatial data warehouse is a subject-oriented, integrated, time-variant, and nonvolatile collection of both spatial and non-spatial data in support of spatial data mining and spatial-data related decision-making processes.

A spatial database stores a large amount of space-related data, such as maps, preprocessed remote sensing or medical imaging data, and VLSI chip layout data.

11.2.1. Characteristics of Spatial Data Mining

- Auto correlation: The properties of locations in a map are often "**auto-correlated**".
- Patterns usually have to be defined in the spatial attribute subspace and not in the complete attribute space.
- Longitude and latitude (or other coordinate systems) are the glue that link different data collections together.
- People are used to maps in GIS; therefore, data mining results have to be summarized on the top of maps.
- Patterns not only refer to points, but can also refer to lines, or polygons or other higher order geometrical objects.
- Large, continuous space defined by spatial attributes.
- Regional knowledge is of particular importance due to lack of global knowledge in geography (→ spatial heterogeneity).

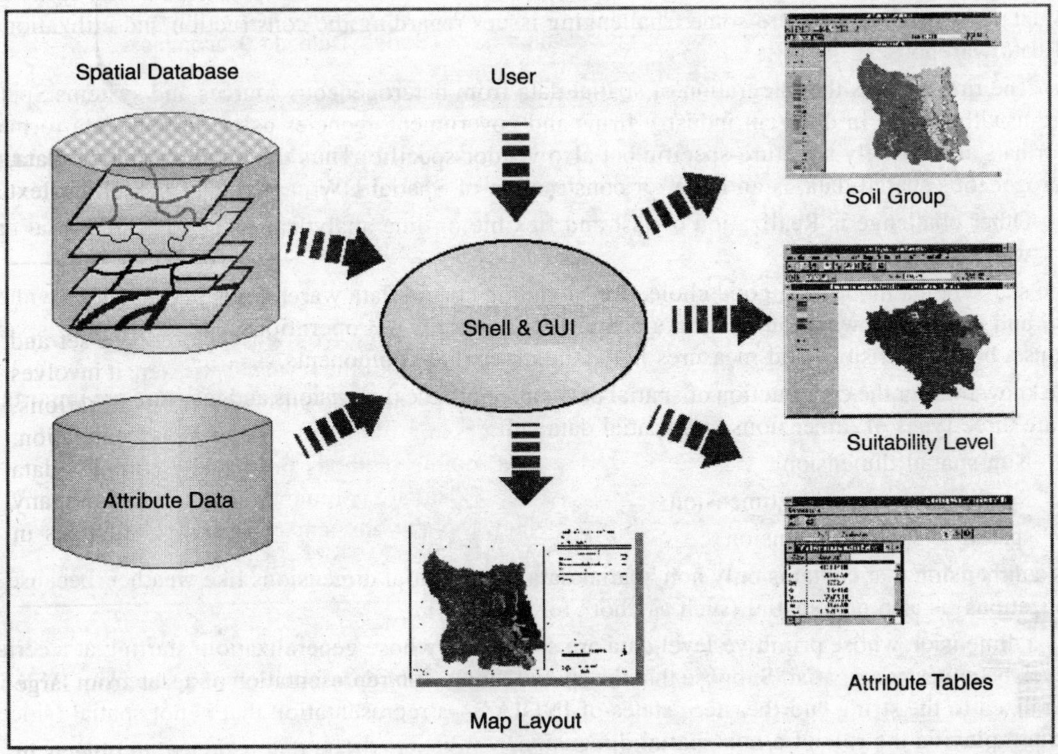

Fig. 11.1. Spatial Database.

11.2.2. Uses of Spatial DM

- Spatial DM is helpful in understanding spatial data.
- It is helpful in discovering spatial relationships and relationships between spatial and non-spatial data.
- Spatial DM can be used for constructing the spatial knowledge bases, reorganizing spatial databases.
- Spatial DM can be used for optimizing spatial queries.

11.2.3. Techniques in Spatial DM **[MDU Dec. 2011; May 2012]**

1. Statistical techniques for spatial data mining:

- Statistical spatial data analysis has been a popular approach to analyzing spatial data and exploring geographic information.
- There is no independence among spatially distributed data because in reality, spatial objects are often interrelated, or more exactly spatially co-located, in the sense that the closer the two objects are located, the more likely they share similar properties.

  For example, climate is likely to be similar in geographically closely located regions. Such a property of close interdependency across nearby space leads to the notion of spatial autocorrelation. Based on this notion, spatial statistical modeling methods have been developed with good success.

2. Spatial Data Cube Construction:

As with relational data, we can integrate spatial data to construct a data warehouse that facilitates spatial data mining. But there are some Challenging issues regarding the construction and utilization of spatial data warehouses:

(*a*) The first issue is the integration of spatial data from heterogeneous sources and systems.Spatial data are usually stored in different industry firms and government agencies using various data formats. Data formats are not only structure-specific but also vendor-specific. Thus the integration and exchange of heterogeneous spatial data, is an issue for construction of Spatial DW.

- Other challenge is Realization of fast and flexible on-line analytical processing in spatial data warehouses.

The star schema model is a good choice for modeling spatial data warehouses because it provides a concise and organized warehouse structure and facilitates OLAP operations. However, in a spatial warehouse, both dimensions and measures may contain spatial components.

We know that for the construction of spatial data cube both the dimensions and measure are important. There are three types of dimensions in a spatial data cube:

1. Non-spatial dimension
2. Spatial-to-non-spatial dimension
3. spatial-to-spatial dimension

Its a dimension that contains only non-spatial data. Non-spatial dimensions like weather because its generalizations is also non-spatial (such as "hot" for weather.)

It's a dimension whose primitive-level data are spatial but whose generalization, starting at a certain high level, becomes non-spatial. Suppose that the spatial dimension representation of a state, say, ASSAM is generalized to the string "northeastern states of INDIA."—a representation that is not spatial (since it is a string), playing the role of a non-spatial dimension.

A spatial-to-spatial dimension is a dimension whose primitive level and all of its high level generalized data are spatial. For example, the dimension equal altitude region contains spatial data, as do all of its generalizations, such as with regions between 0-100 meters above sea level, 101-200meters and so on.

Two types of measure are

1. Numerical measure 2. Spatial measure

A numerical measure contains only numerical data. For example, the monthly rain-fall received in a region, so that a roll-up may compute the total rain-fall received by the region in one year. Numerical measures can be further classified into distributive, algebraic, and holistic.

A spatial measure contains a collection of pointers to spatial objects. For example, in a generalization, the regions with the same range of temperature and rain-fall will be grouped into the same cell, and the measure so formed contains a collection of pointers to those regions.

The computation of spatial measures in spatial data cube construction can be done in 3 ways:

- **No materialization:** do not perform pre-computation of spatial measures in the spatial data cube, only Collection and storage of the corresponding spatial object pointers. .
- **Full materialization:** Pre-compute and store a rough approximation of the spatial measures in the spatial data cube. full materialization leads to extra memory space in order to store all the precomputed cuboids. But one of the solution is to store the precomputed cuboids on secondary storage devices and use them when required.
- **Partial materialization:** Selectively pre-compute some spatial measures in the spatial data cube. As partial means neither full nor no materialization , therefore it presents a trade-off between storage space and response time.

Therefore, it is recommended to perform selection at a finer granularity level. The decision should be based on the utility, share ability of merged regions, and the balanced overall cost of space and on-line computation.

11.2.4. Spatial Clustering Methods

Refer Chapter on Clustering.

11.2.5. Spatial Classification and Spatial Trend Analysis

Spatial Classification: Spatial classification analyzes spatial objects to derive classification schemes in relevance to certain spatial properties, such as the neighborhood of a state, or sea. For example, suppose that you would like to classify regions in a country into hot versus cold. Such classification schemes may be represented in the form of decision trees or rules.

Spatial trend analysis deals with the issue of detection of changes and trends along a spatial dimension. Typically, trend analysis detects changes with time, such as the changes of temporal patterns in time-series data. Spatial trend analysis replaces time with space and studies the trend of non-spatial or spatial data changing with space.

For example, the trend of change in weather when moving towards up in the map. For such analyses, regression and correlation analysis methods are often applied by utilization of spatial data structures and spatial access methods.

Applications

- Spatial data mining is used in geographic information systems.
- Spatial data mining is used in geomarketing.
- Spatial data mining found its application is remote sensing.

Another areas in which spatial Data Mining is used are :

- Image database exploration
- Medical imaging
- Navigation

- Traffic control
- Environmental studies.

11.3. WEB MINING

Web mining is mining of data related to the World Wide Web. This may be the data actually present in Web pages or data related to Web activity. Web data can be:

- Content of actual Web pages
- Intra page structure includes the HTML or XML node for the page.
- Inter page structure is the actual linkage structure between Web pages.
- Usage data that describe how Web pages are accessed by visitors.
- User profiles include demographic and registration information obtained about users.

Fig. 11.2. Web Mining.

NOTE

Web mining is the application of data mining techniques to discover patterns from the web. According to analysis targets, Web mining can be divided into three different types: web usage mining, web content mining, web structure mining.

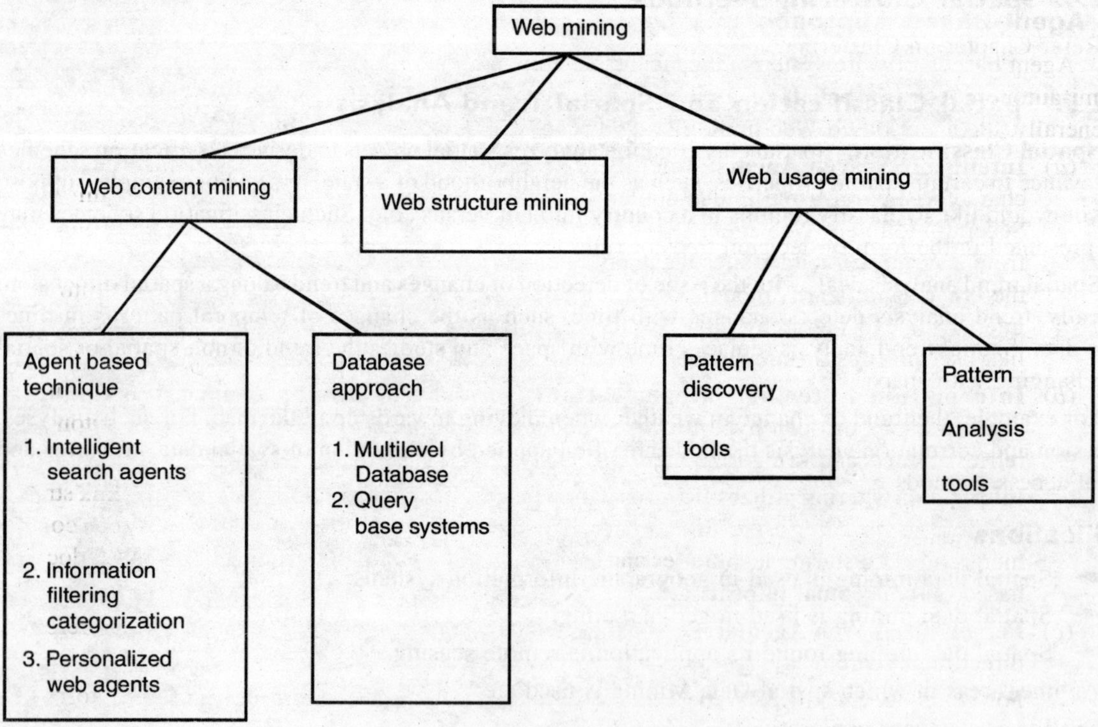

Fig. 11.3. Taxonomy of Web Mining.

Uses of Web Mining

Web mining is used for

- Personalization for a user can be achieved by keeping track of previously accessed pages.
- Web usage patterns can be used to gather business intelligence to improve sales and advertisement.
- Collection of information can be done in new ways.
- Testing of relevance of content and web site architecture can be done with the help of web mining.

11.3.1. Web Content Mining

Web content mining uses data mining techniques for efficiency, effectiveness and scalability.

Web content mining examines the content of Web pages(text +graphics data) as well as results of Web searching. Web content mining is further divided into Web page content mining and web search results mining. Web page content mining is traditional searching of Web pages via content, while Search results mining is a further search of pages found from a previous search. Web content mining is improved version on traditional search engines through techniques such as concept hierarchies and synonyms, user profiles, and analyzing the links between pages.. In recent years these factors have prompted researchers to develop more intelligent tools for information retrieval, such as intelligent Web agents, as well as to extend database and data mining techniques to provide a higher level of organization for semi-structured data available on the Web. These are:

- Agent-based approach
- Database based approach

1. Agent-Based Approach

Agent based(software systems) means the development of AI systems that can act autonomously or semi-autonomously on behalf of a particular user, to discover and organize Web-based information. Generally, the agent-based Web mining systems can be placed into the following three categories.

(*a*) **Intelligent Search Agents:** Intelligent search agents search for relevant information using characteristics of a particular domain to organize and interpret the discovered information. Several intelligent Web agents have been developed like ShopBot retrieves product information from a variety of vendor sites using only general information about the product domain. ILA, on the other hand, learns models of various information sources and translates these into its own internal concept hierarchy. Other example, search engines. Intelligent search agents use techniques like user profiles or knowledge concerning specific domains.

(*b*) **Information Filtering/Categorization:** Various information retrieval techniques and characteristics of open hypertext Web documents are used by the web agents to automatically retrieve, filter, and categorize them.

Information filtering utilizes IR(information retrieval) techniques, knowledge of the link structures to retrieve an categorize documents. For example, BO (Bookmark Organizer) combines hierarchical clustering techniques and user interaction to organize a collection of Web documents based on conceptual information.

(*c*) **Personalized Web Agents:** Personalize Web agents use information about user preferences to direct their search.

For example, Syskill & Webert is a system that utilizes a user profile and learns to rate Web pages of interest using a Bayesian classifier.

2. Database-Based Approach

The database approaches to Web mining have generally focused on techniques for integrating and organizing the heterogeneous and semi-structured data on the Web into more structured and high-level collections of resources, such as in relational databases, and using standard database querying mechanisms and data mining techniques to access and analyze this information.

(*a*) **Multilevel Databases:** The idea behind this approach is that the lowest level of the database contains primitive semi-structured information stored in various Web repositories, such as hypertext documents. At the higher level(s) meta data or generalizations are extracted from lower levels and organized in structured collections such as relational or object-oriented databases.

(*b*) **Web Query Systems:** Web-base query systems and languages are developed to utilize standard database query languages such as SQL, structural information about Web documents, and even natural language processing for accommodating the types of queries that are used in WWW searches. For example, W3QL combines structure queries, based on the organization of hypertext documents, and content queries, based on information retrieval techniques.

11.3.2. Web Usage Mining

Web usage mining performs mining on Web usage data, or Web logs. A Web log is a list of page reference data. It is also called clickstream data because each entry corresponds to a mouse click. and user registration or survey data gathered via tools such as CGI scripts

When these logs are examined from a server perspective, mining uncovers information about the sites where the service resides a can help improve the design of the sites. By evaluating a client's sequence of clicks, information about a user (or a group of users) is detected. This could be used to perform pre-fetching and caching of pages.

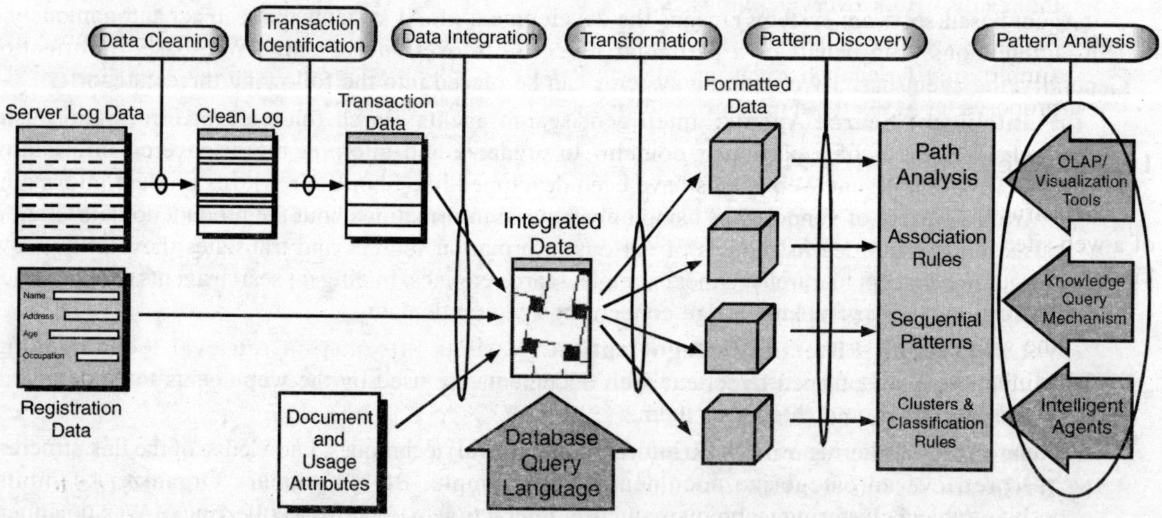

Fig. 11.4. Architecture of Web Usage Mining.

Web usage Mining is used to :

- Personalization for a user can be achieved by keeping track of previously accessed pages.
- By determining frequent access behavior for users, needed links can be identified to improve the performance.

- Identifying common access behaviors can help improve actual design of Web pages and make other modifications to the site.
- Patterns can be used to gather business intelligence to improve sales and advertisement.

More sophisticated systems and techniques for discovery and analysis of patterns can be placed in two main categories.

(a) Pattern Discovery Tools

Pattern discovery tools use techniques from AI, data mining, psychology, and information theory, to mine for knowledge from collected data. For example, the WEBMINER system introduces a general architecture for Web usage mining. WEBMINER automatically discovers association rules and sequential patterns from server access logs.

The most common data mining technique used on clickstream data is that of uncovering traversal pattern. A traversal pattern is a set of pages visited by a user in a session. Other types of patterns may be uncovered by Web usage mining. Different types of Traversal Patterns are:

- Association Rules
- Episodes
- Sequential patterns
- Forward sequences
- Maximal frequent sequences

(b) Pattern Analysis Tools

- Once the patterns have been identifies, analysts need the appropriate tools and techniques to understand, visualize, and interpret these patterns. Some of the generated patterns may be deleted and determined not to be of interest Pattern Analysis is the process of looking at and interpreting the results of discovery activities.
- Sample of pattern analysis tools is: OLAP techniques such as data cubes for the purpose of simplifying the analysis of usage statistics from server access logs. The WEBMINER system proposes an SQL-like query mechanism for querying the discovered knowledge.

11.3.3. Web Structure Mining

Web structure mining is the process of using graph theory to analyze the node and connection structure of a web site. According to the type of web structural data, web structure mining can be divided into two kinds :

1. Extracting patterns from hyperlinks in the web: a hyperlink is a structural component that connects the web page to a different location.
2. Mining the document structure: analysis of the tree-like structure of page structures to describe HTML or XML tag usage.

Structure mining has reduced the two main problems of the WWW.

1. **Irrelevant search results:** Relevance of search information become misconstrued due to the problem that search engines often only allow for low precision criteri
2. The second of these problems is the inability to index the vast amount if information provided on the Web. This causes a low amount of recall with content mining. This minimization comes in part with the function of discovering the model underlying the Web hyperlink structure provided by Web structure mining.

- Techniques of Web Structure Mining.
- Page Rank and HITS Algorithm

Challenges in Web Mining **[MDU Dec. 2011]**

1. The web is dynamic information source that is information is being updated in every few seconds thus the link information and access records are also updated.
2. The web seems to be too large for effective DW and DM. thus it seems to be impossible to set up a large datawarehouse that can store, integrate all kinds of data on the web.
3. The complexity of web page is greater than that of traditional text document collection.
4. Only a small portion of the information available on the web is relevant. As if we want to search something on WWW then we get 1000s of responses but found only some of them useful for us.

11.4. TEXT MINING

Text mining, also known as *text data mining and* refers to the process of deriving high-quality information from text. Text mining usually involves the process of structuring the input text deriving patterns within the structured data, and finally evaluation and interpretating the output.

The objective of Text Mining is to exploit information contained in textual documents in various ways, including …discovery of patterns and trends in data, associations among entities, predictive rules, etc." (Grobelnik et al., 2001)

"Another way to view text data mining is as a process of exploratory data analysis that leads to heretofore unknown information, or to answers for questions for which the answer is not currently known." (Hearst, 1999)

"The non trivial extraction of implicit, previously unknown, and potentially useful information from (large amount of) textual data".

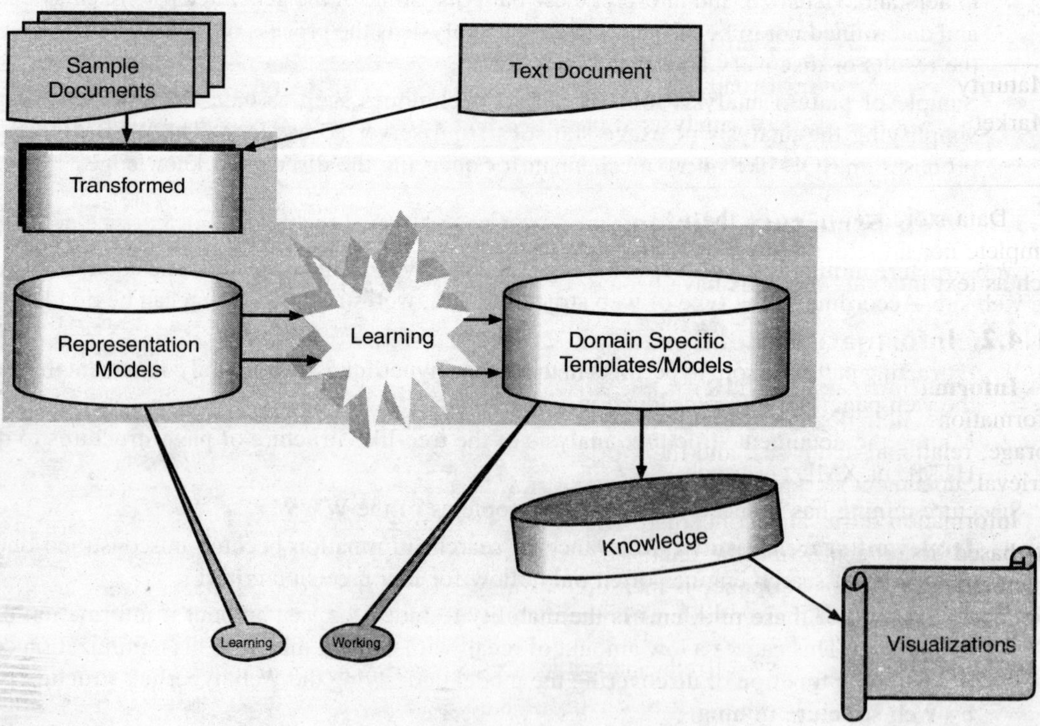

Fig. 11.5. Architecture of Text Mining.

An exploration and analysis of textual (natural-language) data by automatic and semi automatic means to discover new knowledge.

Typical text mining tasks include text categorization, text clustering, concept/entity extraction, production of granular taxonomies, sentiment analysis, document summarization, and entity relation modeling.

Text mining Techniques are:

(*i*) Data mining.
(*ii*) Machine learning.
(*iii*) Information retrieval.
(*iv*) Statistics
(*v*) Natural-language understanding
(*vi*) Case-based reasoning.

11.4.1. Data Mining Vs. Text Mining

	Data Mining	**Text Mining**
Data Object	Numerical and categorical data Data structure Data representation	Textual data StructureUnstructure and semi-structure StraightforwardComplex
Space dimension	< tens of thousands Methods Statistic, neural netowkrs	> tens of thousands Data analysis, machine learning Data mining, information retrieval, NLP, ...
Maturity Market	Broad implementation since 1994 10^5 analysts at large and mid size companies	Broad implementation starting 2000 10^5 analysts corporate workers and individual users

Data stored in most of the text databases are semi structured data it means that data is neither complete nor un structured. Thus to deal with unstructured documents information retrieval techniques such as text indexing methods have been developed.

11.4.2. Information Retrieval

Information retrieval (IR) is the area of study concerned with searching for documents, for information within documents, and for metadata about documents, as well as that of searching structured storage, relational databases, and the World Wide Web. There is overlap in the usage of the terms data retrieval, document retrieval, information retrieval, and text retrieval,

Information retrieval is concerned with organization and retrieval of information from a large set of text based documents while the database systems focused on query and transaction processing of the structured data. The performance is measured using precision and recall. **Precision** is the fraction of the documents retrieved that are relevant to the user's information need.

$$\frac{|\{\text{relevant documents}\} \cap \{\text{retrieved documents}\}|}{|\{\text{retrived documents}\}|}$$

and Recall is the fraction of the documents that are relevant to the query that are successfully retrieved.

$$\text{recall} = \frac{|\{\text{relevant documents}\} \cap \{\text{retrieved documents}\}|}{|\{\text{relevant documents}\}|}$$

Fall-Out is The proportion of non-relevant documents that are retrieved, out of all non-relevant documents available:

$$\text{fall-out} = \frac{|\{\text{non-relevant documents}\} \cap \{\text{retrieved documents}\}|}{|\{\text{non-relevant documents}\}|}$$

Text Retrieval Methods: Retrieval methods fall into two categories: **[MDU May 2012]**

1. Document Selection Methods

- In these methods, the query is regarded as specifying constraints for selecting relevant documents.
- A typical method of this category is the Boolean retrieval model. The retrieval system would take such a Boolean query and return documents that satisfy the Boolean expression.
- The method generally only works well when the user knows a lot about the document collection and can formulate a good query in this way.

2. Document Ranking Methods

- These use the query to rank all documents in the order of relevance. For ordinary users, these methods are more appropriate than document selection methods.
- There are many different ranking methods based on mathematical foundations, including algebra, logic, probability, and statistics.
- The goal is to approximate the degree of relevance of a document with a score computed based on information such as the frequency of words in the document. It is difficult to provide a precise measure of the degree of relevance between a set of keywords.
- Comprehensive empirical evaluation is thus essential for validating any retrieval method.

3. The Most Popular Approach is the Vector Space Model

The basic idea of the model is : We represent a document and a query both as vectors in a high-dimensional space corresponding to all the keywords and use an appropriate similarity measure to compute the similarity between the query vector and the document vector. The similarity values can then be used for ranking documents.

- Tokenization: The first step is to identify keywords for representing documents.
- Stop list: is a set of words that are deemed "irrelevant." To avoid indexing useless words, a text retrieval system often associates a stop list with a set of documents.
- Model a document to facilitate information retrieval:

Starting with a set of d documents and a set of t terms, we can model each document as a vector v in the t dimensional space R^t. Let frequency be the number of occurrences of term t in the document d, that is, freq(d, t).

The (weighted) term-frequency matrix TF(d, t) measures the association of a term with respect to the given document d.

```
TF(d,t) =0 if freq(d, t) =0
= 1+ log(1+log(freq(d,t))) otherwise
```

Inverse document frequency (IDF): Represents the scaling factor, or the importance, of a term t. If a term t occurs in many documents, its importance will be scaled down due to its reduced discriminative power. IDF(t) is defined by the following formula:

IDF(t) = log ((1+|d|)/|d^t|)

where d is the document collection, and dt is the set of documents containing term t. If |dt|<<|d|, the term t will have a large IDF scaling factor and vice versa.

In a complete vector-space model, TF and IDF are combined together, which forms the TF-IDF measure:

TF-IDF(d, t) = TF(d, t) X IDF(t)

11.4.3. Different Text Mining Approaches

1. The major approaches to text mining , based on the inputs taken in the text mining system and the data mining tasks to be performed are:
   - the keyword-based approach, where the input is a set of keywords or terms in the documents,
   - the tagging approach, where the input is a set of tags, and
   - the information-extraction approach, which inputs semantic information, such as events, facts, or entities uncovered by information extraction.
2. Various text mining tasks can be performed on the extracted keywords, tags, or semantic information. These include document clustering, classification, information extraction, association analysis, and trend analysis.
3. Keyword-Based Association Analysis
   - In this analysis sets of keywords or terms that occur frequently together are collected and then the association or correlation relationships among them is found.
   - Association analysis first preprocesses the text data by parsing, stemming, removing stop words, and then evokes association mining algorithms.
   - The association mining process can help detect compound associations, that is, domain-dependent terms or phrases, or non -compound associations. Mining based on these associations is referred to as "term-level association mining".

Advantages

- terms and phrases are automatically tagged so there is no need for human effort in tagging documents;
- the number of meaningless results is greatly reduced, as is the execution time of the mining algorithms.

Document Classification Analysis

Document classification has been used in automated topic tagging , topic directory construction, identification of the document writing styles, and classifying the purposes of hyperlinks. Automated document classification can be performed as follows:

- First, a set of preclassified documents is taken as the training set.
- The training set is then analyzed in order to derive a classification scheme, which is refined with a testing process. The so-derived classification scheme can be used for classification of other on-line documents.
- A few typical classification methods have been used successfully in text classification like nearest-neighbor classification, feature selection methods, Bayesian classification, support vector machines, and association-based classification.
- Association-based classification: classifies documents based on a set of associated, frequently occurring text patterns association-based classification method proceeds as follows:
- First, keywords and terms can be extracted by IR and simple association analysis techniques.

- Second, concept hierarchies of keywords and terms can be obtained using available term classes, such asWordNet, or relying on expert knowledge, or some keyword classification systems.
- A term association mining method can then be applied to discover sets of associated terms that can be used to maximally distinguish one class of documents from others. This derives a set of association rules associated with each document class.

Document Clustering Analysis

The document space is always of very high dimensionality. Due to this, it makes sense to first project the documents into a lower dimensional subspace in which the semantic structure of the document space becomes clear. In the low-dimensional semantic space, the traditional clustering algorithms can then be applied.

Some techniques used are:

- Spectral clustering method: It first performs spectral embedding (dimensionality reduction) on the original data, and then applies the traditional clustering algorithm (*e.g.,* k-means) on the reduced document space.

Drawback

- Can be very expensive to use for very large data sets as they use embedding.
- The mixture model clustering method:

This method models the text data with a mixture model. Clustering involves two steps:

(1) estimating the model parameters based on the text data and any additional prior knowledge, and

(2) inferring the clusters based on the estimated model parameters.

The Latent Semantic Indexing (LSI) and Locality Preserving Indexing (LPI) methods: They are linear dimensionality reduction methods. We can acquire the transformation vectors (embedding function) in LSI and LPI.

LSI seeks to uncover the most representative features rather than the most discriminative features for document representation. Therefore, LSI might not be optimal in discriminating documents with different semantics. LPI aims to discover the local geometrical structure and can have more discriminating power. Compared with LSI and LPI, the PLSI method reveals the latent semantic dimensions in a more interpretable way.

11.4.4. Text Mining Application

The technology is now broadly applied for a wide variety of government, research, and business needs. Other areas where text mining is used are:

1. Security Applications

Many text mining software packages are used for security applications, especially monitoring and analysis of online plain text sources such as Internet news, blogs, etc. for national security purpose. It is also involves in the study of text encryption/decryption.

2. Biomedical Applications

A range of text mining applications are used in the biomedical application. For example PubGene that combines biomedical text mining with network visualization as an Internet service. Text mining is helpful in biomedical application as it Identify relationships and link information from different medical fields.

3. Software Applications

Major firms like IBM, MICROSOFT etc. are conducting research on Text mining methods and software to automate the mining and analysis processes, Within public sector much effort has been concentrated on creating software for tracking and monitoring terrorist activities.

4. Online Media Applications

Text mining is being used by large media companies, such as the Tribune Company, to disambiguate information and to provide readers with greater search experiences, which in turn increases site "stickiness" and revenue.

5. Marketing Applications

Text mining is starting to be used in marketing as well, more specifically in analytical customer relationship management.

6. Academic Applications

Text mining is of great importance to publishers who hold large databases of information needing indexing for retrieval.

Other Applications

Email: Spam filtering

- **News Feeds:** Discover what is interesting
- **Marketing:** Discover distinct groups of potential buyers and make suggestions for other products
- **Industry:** Identifying groups of competitors web pages
- **Job Seeking:** Identify parameters in searching for jobs

11.5. MULTIMEDIA DATABASES MINING

Multimedia is the integration of multiple forms of media. This includes text, graphics, audio, video, etc. CDs and DVDs are often considered to be "multimedia formats" since they can store a lot of data and most forms of multimedia require a lot of disk space.

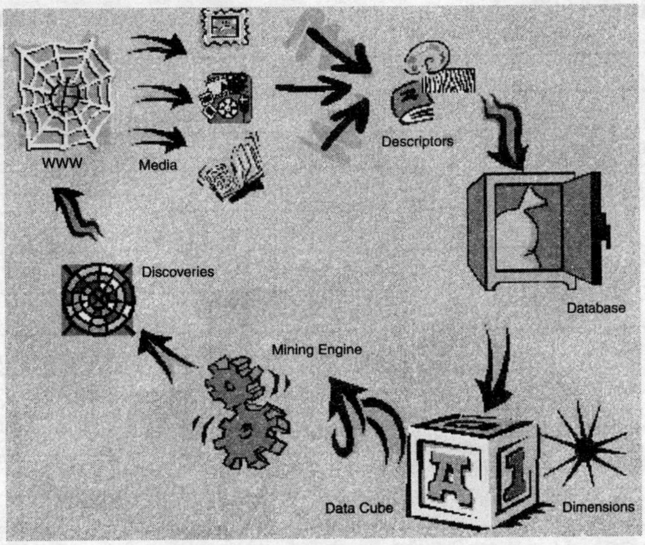

Fig. 11.6. Multimedia Mining.

Multimedia database system stores and manages a large collection of multimedia objects such as audio data, image data, video data hypertext data etc.

Multimedia Mining is discovering knowledge from large amounts of different types of multimedia data. It involves the extraction of implicit knowledge, multimedia data relationships, or other patterns not explicitly stored in multimedia databases.

11.5.1. Multimedia Indexing and Retrieval System

Two types of Multimedia indexing and retrieval system are used for similarity search *i.e.,*, search is either based on data description or data content.

1. Description based retrieval: This is based on building the indices and perform the object retrieval based on image description such as keywords, size, timestamp etc. This method is labor-intensive if performed manually.

And if automated then also results are of poor quality.

2. Content based retrieval: "Content-based" means that the search will analyze the actual contents of the image rather than the metadata such as keywords, tags, and/or descriptions associated with the image. The term 'content' in this context might refer to colors, shapes, textures, or any other information that can be derived from the image itself. CBIR is desirable because most web based image search engines rely purely on metadata.

Multidimensional Analysis of Multimedia Data

Multimedia data cubes can be constructed in the same manner as that for relational data. But multimedia data cube can contain the additional dimensions and measures for multimedia information such as color, texture and shape.

Multimedia database does not store the images rather their descriptors.

(*a*) **Feature descriptor:** It is a set of vectors for each visual characteristic. The main vectors are a

1. Color vector : containing color histogram
2. MFC(most frequent color): Five color centroid
3. MFO(most frequent orientation): Five edge orientation centroid. The edge orientation used are 0°, 45°, 90° etc.

(*b*) **Layout descriptor:** This descriptor contains a color layout vector and an edge layout vector.

Colours	Multivalued	Select 3 most frequent colours
Texture	Multivalued	Select most frequent texture
Format of image	Single Value	
Duration	Single Value	
Size	Single Value	
Width	Single Value	
Height	Single Value	
Keywords	Multivalued	Select before building cube
Image Popularity	Single Value	
Internet domain of image	Single Value	
Internet domain of parent page	Multivalued	Select first parent page
Page Richness	Multivalued	Select first parent page

Fig. 11.7. Dimensions of Multimedia Data.

Multimedia MINER

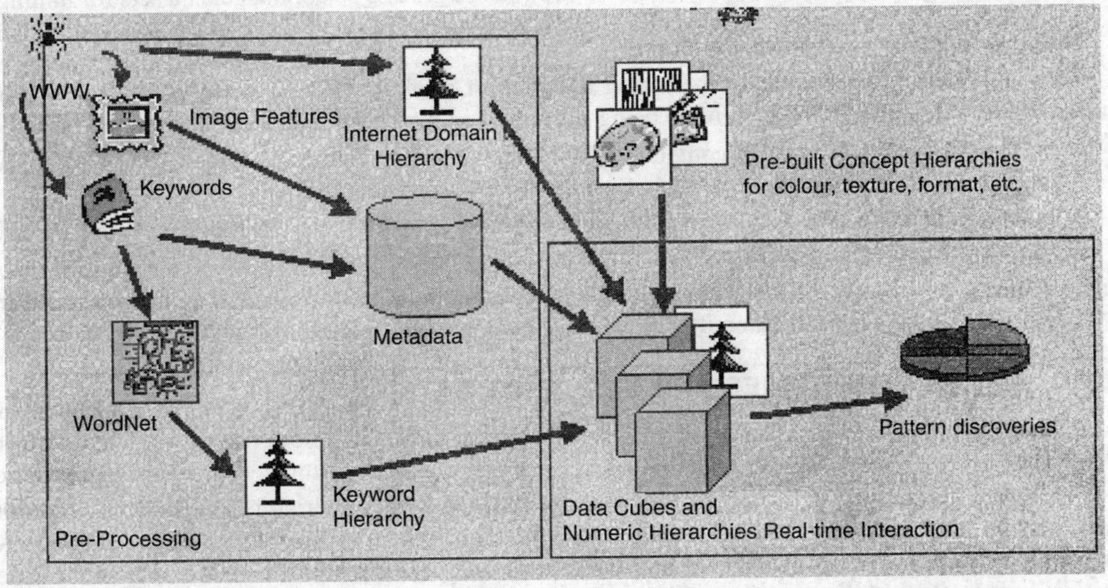

Fig. 11.8. Multimedia Miner.

Multimedia Miner is a system for mining high-level multimedia information and finding resources on the Internet. It is a merge between DBMiner an analytical data mining system, and C-BIRD, a content-based image and video retrieval system.

1. Harvesting: Images and Videos are retrieved from the Internet, alongwith information about the web pages in which they are found.

2. Pre-Processing: Images and Videos are processed to extract colour and texture information. Web pages are processed to extract keywords.

3. Storing: Meta-data extracted is stored in a latabase. This database is used for content-based image and video retrieval.

4. Creating Data Cube: A Multimedia Data Cube is created based on different dimensions (colour, texture, size, width, height, duration, Internet domain, popularity, etc.)

5. Mining: Data Mining algorithms and On-line analytical Processing technology are used to discover implicit knowledge like correlation between multimedia features, classification of images and videos based on their multimedia features, etc.

6. Discovering: Summarization, assocation and classification of multimedia is discovered. licing dicing drilling and rolling-up allow pinpointing specific images and videos by clusters.

7. Image Retrieval: Images and web pages containing these images can be retrieved from the Internet after the data mining process filters out interesting images and videos.

In the pre-processing phase, the keywords extracted from the web pages are "cleaned" and organized in a concept hierarchy.

(*a*) Unnecessary words (stop words) are eliminated;

(*b*) words are normalized (put in their canonical form)

(*c*) WordNet is used to validate words and build the concept hierarchy.

WordNet is a semantic network of 75,000 English words. It gives relationships between meanings of words.

WordNet can also be enriched with new domain specific words that are not in its list.

The built concept hierarchy of words is used to browse images and select image data sets for mining.

Mining association in multimedia data:

There are some difference between mining association rules in MM database vs. in transaction database. Three categories of association is being obsereved in multimedia data, these are:

1. Associations between image content and nonimage content features *e.g.,* if at least 50%of the upper part of water seems to be blue, then it is likely to present the see.
2. Association among image contents that are not related to spatial relationship: *e.g.,* If in a picture there are 2 red rectangles then it is likely to contain one blue circle as well.
3. Associations among the image content related to spatial relationships: *e.g.,* if a blue triangle is in between 2 black squares, then it is likely that there is a big oval shaped object underneath.

11.6. MINING TIME SERIES AND SEQUENCE DATA

Issues and Challenges in Data Mining

Data mining systems face a lot of problems and pitfalls. A system which is quick and correct on some small training sets, could behave completely different when applied to a larger database. A data mining system may work perfect for consistent data and perform significant worse when a little noise is added to the training set. In this section we will discuss the most prominent problems and challenges of data mining systems today.

11.6.1. Noisy and Missing Data

In a large database, many of the attribute values will be inexact or incorrect. This may be due to instruments problem, or human error. Two forms of noise are:

Corrupted Values: Sometimes some of the values in the training set are altered from what they should have been. This may result in one or more tuples in the database conflict with the rules already established. The system may then regard these extreme values as noise, and ignore them. Alternatively, one may take the values into account possibly changing correct patterns recognized. The problem is that one never knows if the extreme values are correct or not, and the challenge is how to handle "weird" values in the best manner.

Missing Attribute Values: One or more of the attribute values may be missing both for examples in the training set and for object which are to be classified. If attributes are missing in the training set, the system may either ignore this object totally, try to take it into account by for instance finding what is the missing attribute's most probable value, or use the value "unknown" as a separate value for the attribute. When an attribute value is missing for an object during classification, the system may check all matching rules and calculate the most probable classification.

Difficult Training Set

Sometimes the training set is not the ultimate training set due to several reasons. These are the following:

Not Representative Data: If the data in the training set is not representative for the objects in the domain, we have a problem. If rules for diagnosing patients are being created and only elderly people are registered in the training set, the result for diagnosing a kid based on these data probably will not be good. Even though this may have serious consequences, we would say that not representative data is mainly a problem of machine learning when the learning is based on few examples. When using large data sets, the rules created probably are representative, as long as the data being classified belongs to the same domain as those in the training set.

No Boundary Cases: To find the real differences between two classes, some boundary cases should be present. If a data mining system for instance is to classify animals, the property counting for a bird might be that it has wings and not that it can fly. This kind of detailed distinction will only be possible if *e.g.,* penguins are registered.

Limited Information: In order to classify an object to a specific class, some condition attributes are investigated. Sometimes, two objects with the same values for condition attributes have a different classification. Then, the objects have some properties which are not among the attributes in the training set, but still make a difference. This is a problem for the system, which does not have any way of distinguish these two types of objects.

Dynamic Nature of Database

We know that the real world databases are large and keep on growing with time. We would like rules which reflect the content of the database at all times, in order to make the best possible classification. Many existing data mining systems require that all the training examples are given at once. If something is changed at a later time, the whole learning process may have to be conducted again. An important challenge for data mining systems is to avoid this, and instead change its current rules according to updates performed.

Databases may be Huge

The size of databases seem to be ever increasing. Most machine learning algorithms have been created for handling only a small training set, for instance a few hundred examples. In order to use similar techniques in databases thousands of times bigger, much care must be taken. Having very much data is advantageous since they probably will show relations really existing, but the number of possible descriptions of such a dataset is enormous. Some possible ways of coping with this problem, are to design algorithms with lower complexity and to use heuristics to find the best classification rules. Simply using a faster computer is seldom a good solution

1. **Security and Privacy:** One of the key issues raised by data mining technology is not a business or technological one, but a social one. It is the issue of individual privacy. Data mining makes it possible to analyze routine business transactions and glean a significant amount of information about individuals buying habits and preferences.
2. **Data Integrity:** Another issue is that of data integrity. Clearly, data analysis can only be as good as the data that is being analyzed. A key implementation challenge is integrating conflicting or redundant data from different sources. For example, a bank may maintain credit cards accounts on several different databases. The addresses (or even the names) of a single cardholder may be different in each. Software must translate data from one system to another and select the address most recently entered.
3. **Cost:** Finally, there is the issue of cost. While system hardware costs have dropped dramatically within the past five years, data mining and data warehousing tend to be self-reinforcing. The more powerful the data mining queries, the greater the utility of the information being gleaned from the data, and the greater the pressure to increase the amount of data being collected and maintained, which increases the pressure for faster, more powerful data mining queries. This increases pressure for larger, faster systems, which are more expensive.

11.7. TRENDS IN DATA MINING

1. **Application exploration:** Since DM systems have some limitations in dealing with application specific DM systems, a trend towards the development of application specific DM systems is being realized these days. Invisible data mining *i.e.,* how mining can be a built-in function is also emerging trend.
2. **Scalable data mining methods:** Since real time data is growing at a very rapid rate therefore DM systems are expected to be able to handle huge amount of data. For this purpose the algorithms for individual and integrated DM functions has become a essential. Also Constraint-based mining that is the use of constraints to guide data mining systems in their search for interesting patterns is a new trend in DM.
3. **Integration with data store systems:** Integration of data mining with database systems, data warehouse systems, and Web database systems is another emerging trend because DW, WWW and the database systems are the most widely used information processing systems.
4. **Standardization of data mining language:** A standard will facilitate systematic development, improve interoperability, and promote the education and use of data mining systems in industry and society.
5. **New methods for mining complex types of data:** More research is required towards the integration of data mining methods with existing data analysis techniques for the complex types of data.
6. **Web mining:** As web is playing an important role in today's business, web content mining, web usage mining and web structure mining, thus becoming one of the crucial area in DM.
7. **Privacy protection and information security in data mining:** in addition to provide the data access, the techniques to ensure privacy protection and information security must also be developed.

11.8. MINING TIME SERIES DATA

A time series is a collection of observations of well-defined data items obtained through repeated measurements over time. For example, measuring the value of retail sales each month of the year would comprise a time series. This is because sales revenue is well defined, and consistently measured at equally spaced intervals. Data collected irregularly or only once are not time series

Examples: Weather data, Climate data, Seismic waves, Financial markets, medical treatments, production systems etc.

Time series data may be continuous (observations can be recorded at every instant of time) like ECG or discrete(observations can be recorded at spaced interval of time) like rainfall etc.

An observed time series can be decomposed into three components: the trend (long term direction), the seasonal (systematic, calendar related movements) and the irregular (unsystematic, short term fluctuations) and cyclic.

1. Trend Component: The trend is defined as the 'long term' movement in a time series without calendar related and irregular effects, and is a reflection of the underlying level. It is the result of influences such as population growth, price inflation and general economic changes. The methods for determining the trend line is to take the average over a certain period of time.

➡ The following graph depicts a series in which there is an obvious upward trend over time:

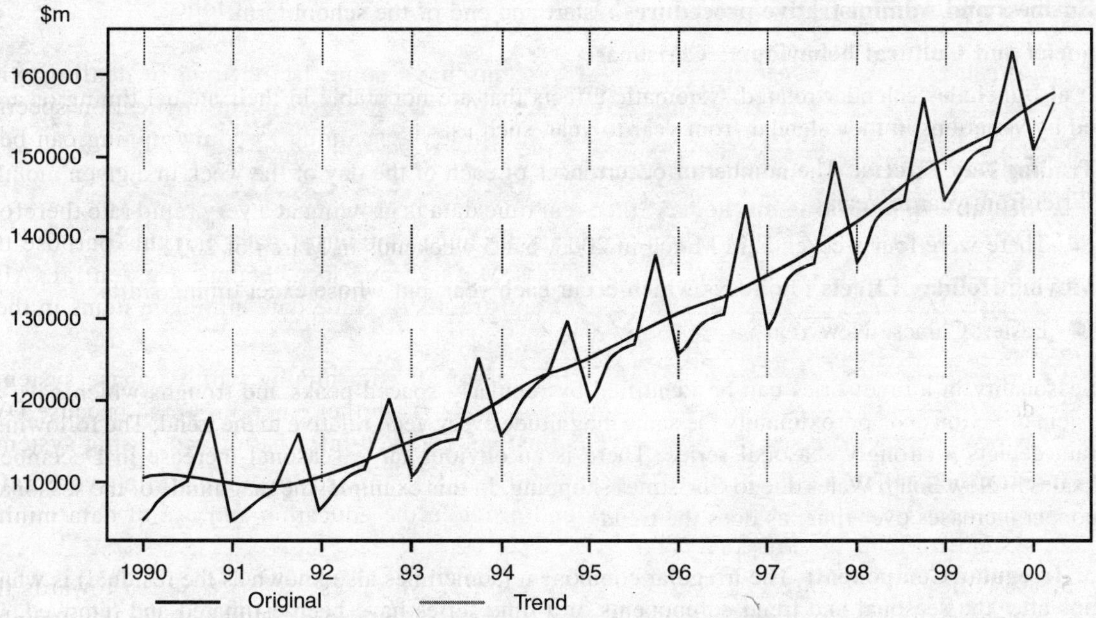

Fig. 11.9. Quarterly Gross Domestic Product.

2. Seasonal Component: The seasonal component consists of effects that are reasonably stable with respect to timing, direction and magnitude. It arises from systematic, calendar related influences such as:

Natural Conditions : Weather fluctuations that are representative of the season (uncharacteristic weather patterns such as snow in summer would be considered irregular influences).

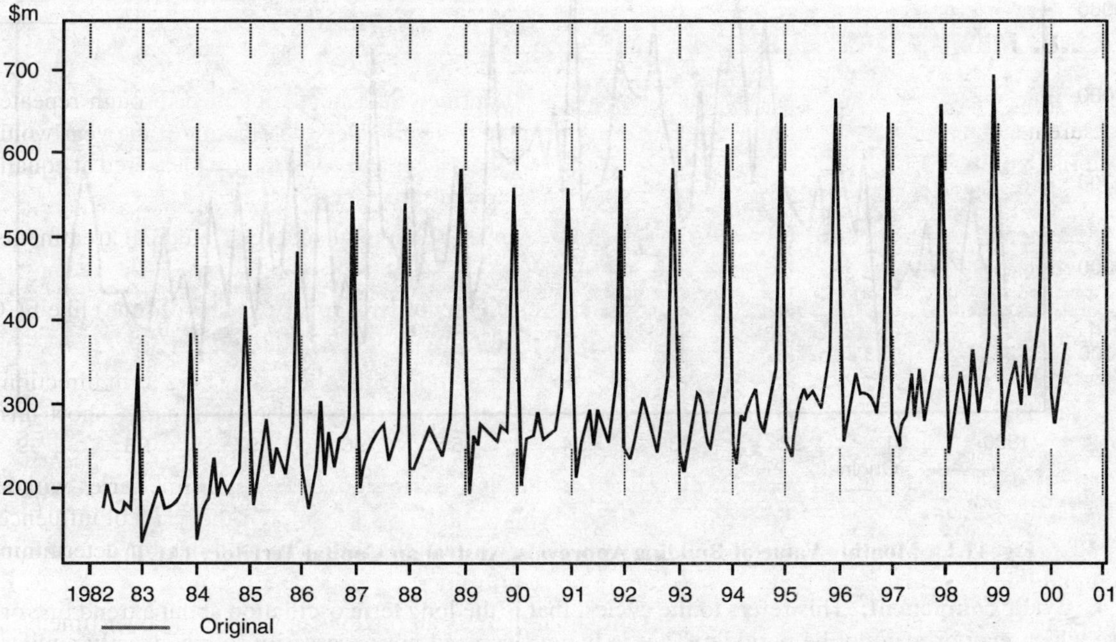

Fig. 11.10. Monthly Retail Sales in New South Wales (NSW) Retail Department Stores.

Business and Administrative procedures : start and end of the school term.

Social and Cultural behaviour : Christmas.

It also includes calendar related systematic effects that are not stable in their annual timing or are caused by variations in the calendar from year to year, such as:

Trading Day Effects: The number of occurrences of each of the day of the week in a given month will differ from year to year.

- There were four weekends in March in 2000, but 5 weekends in March of 2002.

Moving Holiday Effects : holidays which occur each year, but whose exact timing shifts.

- Easter, Chinese New Year.

Seasonality in a time series can be identified by regularly spaced peaks and troughs which have a consistent direction and approximately the same magnitude every year, relative to the trend. The following diagram depicts a strongly seasonal series. There is an obvious large seasonal increase in December retail sales in New South Wales due to Christmas shopping. In this example, the magnitude of the seasonal component increases over time, as does the trend.

3. Irregular Component: The irregular component (sometimes also known as the residual) is what remains after the seasonal and trend components of a time series have been estimated and removed. It results from short term fluctuations in the series which are neither systematic nor predictable. In a highly irregular series, these fluctuations can dominate movements, which will mask the trend and seasonality. The following graph is of a highly irregular time series:

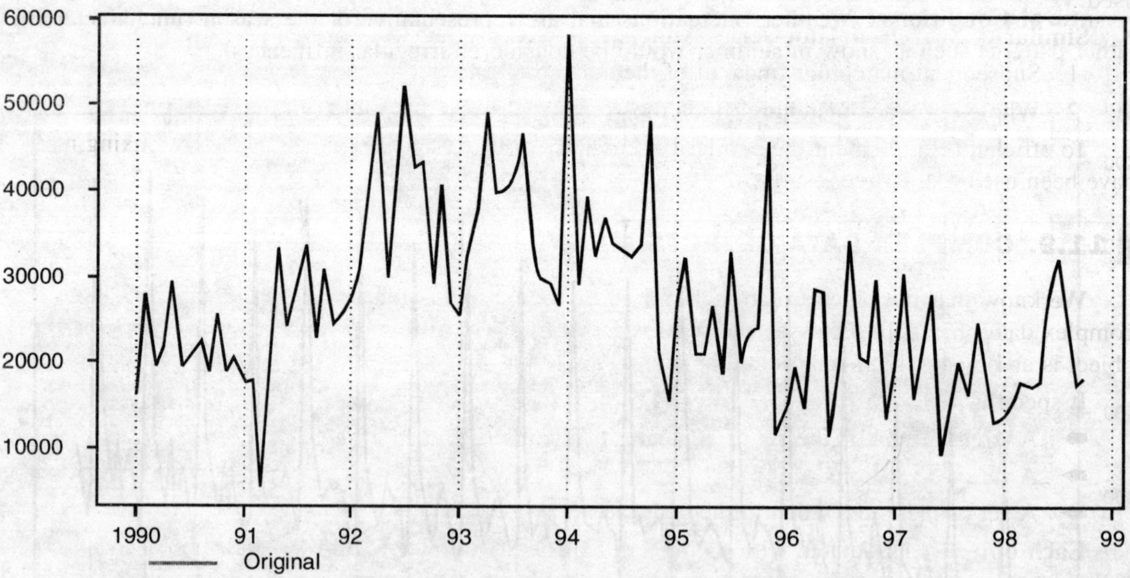

Fig. 11.11. Monthly Value of Building Approvals, Australian Capital Territory (ACT)

4. Cyclic component: This refers to the cycles, that is the long term oscillation about a trend line or curve which may or may not be periodic. That is the cycles need not necessarily follow exactly similar patterns after equal intervals of time.

11.8.1. Two Aspects of Time Series Analysis

Trend Analysis: There are no proven "automatic" techniques to identify trend components in the time series data; however, as long as the trend is monotonous (consistently increasing or decreasing) that part of data analysis is typically not very difficult. If the time series data contain considerable error, then the first step in the process of trend identification is smoothing.

Smoothing. Smoothing always involves some form of local averaging of data such that the nonsystematic components of individual observations cancel each other out. The most common technique is *moving average* smoothing which replaces each element of the series by either the simple or weighted average of *n* surrounding elements, where *n* is the width of the smoothing "window" (see Box & Jenkins, 1976; Velleman & Hoaglin, 1981). Medians can be used instead of means. The main advantage of median as compared to moving average smoothing is that its results are less biased by outliers (within the smoothing window). Thus, if there are outliers in the data (*e.g.,* due to measurement errors), median smoothing typically produces smoother or at least more "reliable" curves than moving average based on the same window width. The main disadvantage of median smoothing is that in the absence of clear outliers it may produce more "jagged" curves than moving average and it does not allow for weighting.

In the relatively less common cases (in time series data), when the measurement error is very large, the *distance weighted least squares smoothing* or *negative exponentially weighted smoothing* techniques can be used. All those methods will filter out the noise and convert the data into a smooth curve that is relatively unbiased by outliers (see the respective sections on each of those methods for more details). Series with relatively few and systematically distributed points can be smoothed with *bicubic splines.*

Fitting a function. Many monotonous time series data can be adequately approximated by a linear function; if there is a clear monotonous nonlinear component, the data first need to be transformed to remove the nonlinearity. Usually a logarithmic, exponential, or (less often) polynomial function can be used.

Similarity Search in Time-series Analysis: There are two types of similarity search:

1. Subsequent matching: finds all of the data similar to the given sequence.
2. Whole sequence matching: finds those sequences that are similar to one another.

To efficiently implement the similarity search in large database, various kinds of indexing methods have been used like R-trees, suffix trees etc.

11.9. COMPLEX DATA OBJECTS MINING

We know that the object relational and object – oriented database systems organize a large set of complex data objects into classes and classes are again organized into subclasses or hierarchies. A class object is an object which acts as a template.

It specifies:

- A structure that is the set of attributes of the instances.
- A set of operations.
- A set of methods which implement the operations.

Each object is associated with an object identifier, set of attributes (set-valued or list-valued data) and a set of methods (specify the rules associated with the object class).

Generalization in object-relational and object oriented databases.

Generalization is the process of extracting shared characteristics from two or more classes, and combining them into a generalized superclass. Shared characteristics can be attributes, associations, or methods :

1. *Set-valued attributes:* May be homogeneous or non homogeneous type. Set-valued attributes may be generalized to Generalization of each value in the set into its corresponding higher-level concepts.

Derivation of the general behavior of the set, such as the number of elements in the set, the types or value ranges in the set, or the weighted average for numerical data

For example, *hobby* = {*badminton, guitar, making food*} generalizes to {*sports, music, cooking*}.

2. List-valued or a sequence-valued attribute: Same as set-valued attributes except that the order of the elements in the sequence should be observed in the generalization

Aggregation and Approximation in Spatial and Multimedia Data Generalization

Aggregation and approximation are another important means of generalization. They are especially useful for generalizing attributes with large sets of values, complex structures, and spatial or multimedia data.

Example: Spatial aggregation and approximation. Suppose that we have different pieces of land for various purposes of agricultural usage, such as the planting of vegetables, grains, and fruits. These pieces can be merged or aggregated into one large piece of agricultural land by a spatial merge. However, such a piece of agricultural land may contain highways, houses, and small stores. If the majority of the land is used for agriculture, the scattered regions for other purposes can be ignored, and the whole region can be claimed as an agricultural area by approximation

11.9.1. Generalization of Object Identifiers and Class/Subclass Hierarchies

An object identifier can be generalized as follows. First, the object identifier is generalized to the identifier of the lowest subclass to which the object belongs. The identifier of this subclass can then, in turn, be generalized to a higherlevel class/subclass identifier by climbing up the class/subclass hierarchy. Similarly, a class or a subclass can be generalized to its corresponding superclass(es) by climbing up its associated class/subclass hierarchy.

Generalization of Class Composition Hierarchies

An attribute of an object may be composed of or described by another object, some of whose attributes may be in turn composed of or described by other objects, thus forming a class composition hierarchy. Generalization on a class composition hierarchy can be viewed as generalization on a set of nested structured data (which are possibly infinite, if the nesting is recursive).

Construction and Mining of Object Cubes

In an object database, data generalization and multidimensional analysis are not applied to individual objects but to classes of objects. Since a set of objects in a class may share many attributes and methods, and the generalization of each attribute and method may apply a sequence of generalization operators, the major issue becomes how to make the generalization processes cooperate among different attributes and methods in the class(es).

Generalization-Based Mining of Plan Databases by Divide-and-Conquer

A plan consists of a variable sequence of actions. A plan database, or simply a planbase, is a large collection of plans. Plan mining is the task of mining significant patterns or knowledge from a planbase. Plan mining can be used to discover the travel patterns of the passengers in air flight database. On the other hand in sequential pattern mining a large number of frequently occurring sequences are mined at a very detailed level.

Meaning of Divide-and-Conquer : Divide-and-Conquer strategies finds the high level summary of sequence of plans by multidimensional generalization of a planbase and then divides the planbase based on mined patterns to discover the corresponding characteristics of subplanbases.

11.10. SEQUENCE PATTERN MINING

Sequential pattern mining is the mining of frequently occurring ordered events or subsequences as patterns.

An example of a sequential pattern is "Customers who buy a Canon digital camera are likely to buy an HP color printer within a month.

Applications: Sequence pattern data are the time sequence data therefore it found its importance in

- shelf placement and promotions in retail.
- telecommunications and other businesses, for targeted marketing, customer retention.
- Web access pattern analysis,
- weather prediction,
- production processes,
- and network intrusion detection

Sequential pattern mining is done on symbolic patterns, whereas numerical curve analysis usually belongs to the scope of trend analysis and forecasting in statistical time-series analysis (explained in time series analysis topic).

11.10.1. Parameters for Sequential Pattern Mining

Certain parameters which affect the results of sequential pattern mining are:

1. duration of time sequence (t): the duration may be the available sequence in a database or user may select it like in SQL "where year=2000".
2. Event folding window (w): A set of events occurring within a specified period of time can be viewed as occurring together in certain analysis.
3. Time interval: time interval (int) is the interval between the events in the discovered pattern. If int=0, this means that no interval gap is allowed.

11.10.2. Approaches for Sequential Pattern Mining

- Apriori-based Approaches(explained in chapter association rules)
  - GSP
  - SPADE
- Pattern-Growth-based Approaches
  - FreeSpan
  - PrefixSpan
- GSP(Generalized Sequential Patterns) adopts a candidate generate-and-test approach using horizonal data format (where the data are represented as <sequence ID : sequence_of_itemsets>.
- SPADE (Sequential PAttern Discovery using Equivalent classes) adopts a candidate generate and test approach using vertical data format (where the data are represented as <itemset :(sequence ID, event ID)>. The vertical data format can be obtained by transforming from a horizontally formatted sequence database in just one scan.
- PrefixSpan is a pattern growth method, which does not require candidate generation.

Challenges on Sequential Pattern Mining

- A huge number of possible sequential patterns are hidden in databases

➡ A mining algorithm should
- find the complete set of patterns, when possible, satisfying the minimum support (frequency) threshold.
- be highly efficient, scalable, involving only a small number of database scans.
- be able to incorporate various kinds of user-specific constraints.

Summary

- This chapter is about mining the complex types of data including the object data, spatial data, time-series data, web data, text data etc.
- There is a difference between Transaction databases vs. sequence databases

A transaction database

TID	Itemsets
10	a, b, d
20	a, c, d
30	a, d, e
40	b, e, f

A sequence database

SID	sequences
10	<a(abc)(ac)d(cf)>
20	<(ad)c(bc)(ae)>
30	<(ef)(ab)(df)cb)>
40	<eg(af)cbc>

- Spatial data mining refers to the extraction of knowledge, spatial relationships, or other interesting patterns not explicitly stored in spatial databases. Such mining demands an integration of data mining with spatial database technologies and A spatial data warehouse is a subject-oriented, integrated, time-variant, and nonvolatile collection of both spatial and non-spatial data in support of spatial data mining and spatial-data related decision-making processes
- Text Mining is the procedure of synthesizing information, by analyzing relations, patterns, and rules among textual data. Procedures in Text Mining:
  - ➡ Text Summarization: to extract its partial content reflection its whole contents automatically.
  - ➡ Text Categorization: to assign a category to the text among categories predefined by users.
  - ➡ Text clustering: to segment texts into several clusters, depending on the substantial relevance.
- Web mining is mining of data related to the World Wide Web. This may be the data actually present in Web pages or data related to Web activity. Web data can be:
  - ➡ Content of actual Web pages
  - ➡ Intra page structure includes the HTML or XML node for the page.
  - ➡ Inter page structure is the actual linkage structure between Web pages.
  - ➡ Usage data that describe how Web pages are accessed by visitors.
  - ➡ User profiles include demographic and registration information obtained about users.
- Multimedia data mining is the discovery of interesting pattern from MM databases that store and manage large collection of MM objects like video, audio, image, hypertext data etc.
- A time-series database consists of sequence of values or events changing with time.

Review Questions

1. Explain Challenges in Text Mining.

   Hint: Data collection is "free text"

   - Data is not well-organized
   - Semi-structured or unstructured
   - Natural language text contains ambiguities on many levels
   - Lexical, syntactic, semantic, and pragmatic
   - Learning techniques for processing text typically need annotated training examples
   - Consider bootstrapping techniques

2. Discuss some challenges in data mining.
3. List the Activities of Web usage Mining.

   HINT:

   - Preprocessing activities center around reformatting the Web log data before processing.
   - Pattern discovery activities form the major portion of the mining activities because these activities look to find hidden patterns within the log data.
   - Pattern analysis is the process of looking at and interpreting the results of the discovery activities.

   Preprocessing:

   - It includes cleansing, user identification , session identification, path completion and formatting.
   - Before processing the log, the data may be changed in several ways. For security or privacy reasons, the page addresses may be change into unique page identifications. This also will save storage space. Also, the data may be cleansed by removing irrelevant information.
   - All pages visited from one source could be grouped by a server to understand the patterns of page references from each user.
   - A common technique is for a server site to divide the log records into sessions. A session is a set of page references from one source site during one logical period. Login and logoff represents the logical start and end of the session.

4. What do u mean by Data Structures, explain suffix tree ?

   - Data Structures are used to keep track of patterns identified during the Web usage mining process.
   - One of the basic data structure is trie. A trie is a rooted tree, where each path from the root to leaf represents a sequence. They are used to store strings for pattern-matching applications.
   - The wastage of space in trie is solved by compressing nodes together when they have degrees of one. A compressed trie is called suffix tree.

5. Why Page Rank is used ?

   Hint:

   - Was designed to increase the effectiveness and improve the efficiency of the search engines.
   - It is used to measure the importance of a page and to prioritize the pages returned from a traditional search engine using keyword searching.
   - Google uses this technique.
   - The PageRank value for a page is calculated based on the number of pages that point to it

6. Explain Virtual Web View.

   Hint:

   1. Multiple layered database (MLDB) is used to handle large amounts of unstructured data on the Web.
   2. This database is massive and distributed. Each layer is more generalized than the layer beneath it.

3. The MLDB provides an abstracted and condensed view of a portion of the Web. A view of the MLDB, which is called a Virtual Web View (VWV) can be constructed.
4. Generalization tools are proposed, and concept hierarchies are used to assist in the generalization process for constructing the higher levels of the MLDB.
5. WebML, a web data mining query language is proposed to provide data mining operations on the MLDB. It is an extension of DMQL.

7. What do you mean by Harvest System?

1. The Harvest system is based on the use of caching, indexing, and crawling. Harvest is actually a set of tools that facilitate gathering of information from diverse sources.
2. The Harvest design is centered around the use of gatherers and brokers.
3. A gatherer obtains information for indexing from an Internet service provider, while a broker provides the index and query interface. The relationship between brokers and gatherers can vary. Brokers may interface directly with gatherers or may go through other brokers to get to the gatherers.
4. Indices in Harvest are topic-specific, as are brokers.
5. Harvest gatherers use Essence system to assist in collecting data, Essence classifies documents by creating a semantic index.
6. Semantic indexing generates different types of information for different types of files. It then creates indices on this information.

8. What are stock and flow series ?

Hint: Time series can be classified into two different types: stock and flow.

A stock series is a measure of certain attributes at a point in time and can be thought of as "stocktakes". For example, the **Monthly Labour Force Survey** is a stock measure because it takes stock of whether a person was employed in the reference week.

Flow series are series which are a measure of activity over a given period. For example, surveys of **Retail Trade** activity. Manufacturing is also a flow measure because a certain amount is produced each day, and then these amounts are summed to give a total value for production for a given reporting period

The main difference between a stock and a flow series is that flow series can contain effects related to the calendar (trading day effects). Both types of series can still be seasonally adjusted using the same seasonal adjustment process.

❑❑❑

QUESTION BANK

1. Define Data mining.

Ans. It refers to extracting or "mining" knowledge from large amount of data. Data mining is a process of discovering interesting knowledge from large amounts of data stored either, in database, data warehouse, or other information repositories.

2. Give some alternative terms for data mining.

Ans.
- Knowledge mining
- Knowledge extraction
- Data/pattern analysis.
- Data Archaeology
- Data dredging

3. What is KDD.

Ans. KDD-Knowledge Discovery in Databases.

4. What are the steps involved in KDD process.

Ans.
- Data cleaning
- Data Mining
- Pattern Evaluation
- Knowledge Presentation
- Data Integration
- Data Selection
- Data Transformation.

5. What is the use of the knowledge base?

Ans. Knowledge base is domain knowledge that is used to guide search or evaluate the interestingness of resulting pattern. Such knowledge can include concept hierarchies used to organize attribute/ attribute values in to different levels of abstraction.

6. Arcitecture of a typical data mining system.

Ans.

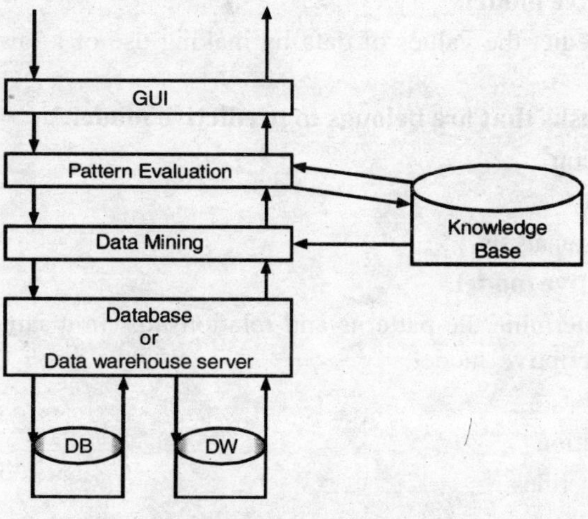

Fig. 1. Data Mining Architecture

7. Mention some of the data mining techniques.

Ans.
- Statistics
- Machine learning
- Decision Tree
- Hidden markov models
- Artificial Intelligence
- Genetic Algorithm
- Meta learning

8. Give few statistical techniques.

Ans.
- Point Estimation
- Data Summarization
- Bayesian Techniques
- Testing Hypothesis
- Correlation
- Regression

9. What is meta learning.

Ans. Concept of combining the predictions made from multiple models of data mining and analyzing those predictions to formulate a new and previously unknown prediction.
- GUI
- Pattern Evaluation
- Database or Data warehouse server
- DB DW

10. Define Genetic algorithm.

Ans.
- Search algorithm.
- Enables us to locate optimal binary string by processing an initial random population of binary strings by performing operations such as artificial mutation, crossover and selection.

11. What is the purpose of Data mining Technique?

Ans. It provides a way to use various data mining tasks.

12. Define Predictive model.

Ans. It is used to predict the values of data by making use of known results from a different set of sample data.

13. Data mining tasks that are belongs to predictive model.

Ans.
- Classification
- Regression
- Time series analysis.

14. Define descriptive model.

Ans. It is used to determine the patterns and relationships in a sample data. Data mining tasks that belongs to descriptive model:
- Clustering
- Summarization
- Association rules
- Sequence discovery

15. Define the term summarization.

Ans. The summarization of a large chunk of data contained in a web page or a document.

Summarization = charcterization = generalization

16. List out the advanced database systems.

Ans.
- Extended-relational databases
- Object-oriented databases
- Deductive databases
- Spatial databases
- Temporal databases
- Multimedia databases
- Active databases
- Scientific databases
- Knowledge databases

17. Define cluster analysis.

Ans. Cluster analyses data objects without consulting a known class label. The class labels are not present in the training data simply because they are not known to begin with.

18. Classifications of Data mining systems.

Ans.
- Based on the kinds of databases mined:
  - ■ According to model
- Relational mining system
- Transactional mining system
- Object-oriented mining system
- Object-Relational mining system
- Data warehouse mining system

Types of Data:
- Spatial data mining system
- Time series data mining system
- Text data mining system
- Multimedia data mining system
- Based on kinds of Knowledge mined

According to functionalities:
- Characterization
- Discrimination
- Association
- Classification
- Clustering
- Outlier analysis
- Evolution analysis

According to levels of abstraction of the knowledge mined:
- Generalized knowledge (High level of abstraction)
- Primitive-level knowledge (Raw data level)

According to mine data regularities versus mine data irregularities:

- Based on kinds of techniques utilized

According to user interaction:

- Autonomous systems
- Interactive exploratory system
- Query-driven systems

According to methods of data analysis:

- Database-oriented
- Data warehouse-oriented
- Machine learning
- Statistics
- Visualization
- Pattern recognition
- Neural networks
- Based on applications adopted

Finance
Telecommunication
DNA
Stock markets
E-mail and so on

19. Describe challenges to data mining regarding data mining methodology and user interaction issues.

Ans.
- Mining different kinds of knowledge in databases
- Interactive mining of knowledge at multiple levels of abstraction
- Incorporation of background knowledge
- Data mining query languages and ad hoc data mining
- Presentation and visualization of data mining results
- Handling noisy or incomplete data
- Pattern evaluation.

20. Describe challenges to data mining regarding performance issues.

Ans.
- Efficiency and scalability of data mining algorithms
- Parallel, distributed, and incremental mining algorithms.

21. Describe issues relating to the diversity of database types.

Ans.
- Handling of relational and complex types of data.
- Mining information from heterogeneous databases and global information systems.

22. What is meant by pattern?

Ans. Pattern represents knowledge if it is easily understood by humans; valid on test data with some degree of certainty; and potentially useful, novel,or validates a hunch about which the used was curious. Measures of pattern interestingness, either objective or subjective, can be used to guide the discovery process.

23. How is a data warehouse different from a database?

Ans. Data warehouse is a repository of multiple heterogeneous data sources, organized under a unified schema at a single site in order to facilitate management decision-making. Database consists of a collection of interrelated data.

24. Define Association Rule Mining.

Ans. Association rule mining searches for interesting relationships among items in a given data set.

25. When we can say the association rules are interesting?

Ans. Association rules are considered interesting if they satisfy both a minimum support threshold and a minimum confidence threshold. Users or domain experts can set such thresholds.

26. Explain Association rule in mathematical notations.

Ans. Let I-{i1, i2, ..., im} be a set of items

Let D, the task relevant data be a set of database transaction T is a set of items.

An association rule is an implication of the form A=>B where A C I, B C I,

and An B=ϕ. The rule A=>B contains in the transaction set D with support s, where s is the percentage of transactions in D that contain A ∪ B. The Rule A=> B has confidence c in the transaction set D if c is the percentage of transactions in D containing A that also contain B.

27. Define support and confidence in Association rule mining.

Ans. Support S is the percentage of transactions in D that contain A ∪ B. Confidence c is the percentage of transactions in D containing A that also contain B.

Support (A=>B)= P(A ∪ B)

Confidence (A=>B)=P(B/A)

28. How are association rules mined from large databases?

Ans. **Step I:** Find all frequent item sets.

Step II: Generate strong association rules from frequent item sets.

29. Describe the different classifications of Association rule mining.

Ans. Based on types of values handled in the Rule

(*i*) Boolean association rule

(*ii*) Quantitative association rule

Based on the dimensions of data involved

(*i*) Single dimensional association rule

(*ii*) Multidimensional association rule

Based on the levels of abstraction involved

(*i*) Multilevel association rule

(*ii*) Single level association rule

Based on various extensions

(*i*) Correlation analysis

(*ii*) Mining max patterns

30. What is the purpose of Apriori Algorithm?

Ans. Apriori algorithm is an influential algorithm for mining frequent item sets for Boolean association rules. The name of the algorithm is based on the fact that the algorithm uses prior knowledge of frequent item set properties.

31. Define anti-monotone property.

Ans. If a set cannot pass a test, all of its supersets will fail the same test as well.

32. How to generate association rules from frequent item sets?

Ans. Association rules can be generated as follows:

For each frequent item set1, generate all non empty subsets of 1.

For every non empty subsets s of 1, output the rule "S=>(1-s)"if

$$\frac{\text{Support count(1)}}{\text{Support\_count(s)}} = \text{min\_conf},$$

Where min_conf is the minimum confidence threshold.

33. Give few techniques to improve the efficiency of Apriori algorithm.

Ans.
- Hash based technique
- Transaction Reduction
- Portioning
- Sampling
- Dynamic item counting.

34. What are the things suffering the performance of Apriori candidate generation technique.

Ans.
- Need to generate a huge number of candidate sets.
- Need to repeatedly scan the scan the database and check a large set of candidates by pattern matching.

35. Describe the method of generating frequent item sets without candidate generation.

Ans. Frequent-pattern growth(or FP Growth) adopts divide-and-conquer strategy.

Steps: Compress the database representing frequent items into a frequent pattern tree or FP tree. Divide the compressed database into a set of conditional database. Mine each conditional database separately.

36. Define Iceberg query.

Ans. It computes an aggregate function over an attribute or set of attributes in order to find aggregate values above some specified threshold. Given relation R with attributes a1,a2,…..,an and b, and an aggregate function,

```
agg_f, an iceberg query is the form
Select R.a1,R.a2,.....R.an,agg_f(R,b)
From relation R
Group by R.a1,R.a2,....,R.an
Having agg_f(R.b)>=threhold
```

37. Mention few approaches to mining Multilevel Association Rules

Ans.
- Uniform minimum support for all levels(or uniform support)
- Using reduced minimum support at lower levels(or reduced support)
- Level-by-level independent
- Level-cross filtering by single item
- Level-cross filtering by k-item set

38. What are multidimensional association rules?

Ans. Association rules that involve two or more dimensions or predicates

- **Interdimension association rule:** Multidimensional association rule with no repeated predicate or dimension.
- **Hybrid-dimension association rule:** Multidimensional association rule with multiple occurrences of some predicates or dimensions.

39. Define constraint-Based Association Mining.

Ans. Mining is performed under the guidance of various kinds of constraints provided by the user. The constraints include the following

- Knowledge type constraints
- Data constraints
- Dimension/level constraints
- Interestingness constraints
- Rule constraints.

40. Define the concept of classification.

Ans. Two step process:

- A model is built describing a predefined set of data classes or concepts. The model is constructed by analyzing database tuples described by attributes.
- The model is used for classification.

41. What is Decision tree?

Ans. A decision tree is a flow chart like tree structures, where each internal node denotes a test on an attribute, each branch represents an outcome of the test, and leaf nodes represent classes or class distributions. The top most in a tree is the root node.

42. What is Attribute Selection Measure?

Ans. The information Gain measure is used to select the test attribute at each node in the decision tree. Such a measure is referred to as an attribute selection measure or a measure of the goodness of split.

43. Describe Tree pruning methods.

Ans. When a decision tree is built, many of the branches will reflect anomalies in the training data due to noise or outlier. Tree pruning methods address this problem of over fitting the data.

Approaches :

- Pre pruning
- Post pruning

44. Define Pre Pruning.

Ans. A tree is pruned by halting its construction early. Upon halting, the node becomes a leaf. The leaf may hold the most frequent class among the subset samples.

45. Define Post Pruning.

Ans. Post pruning removes branches from a "Fully grown" tree. A tree node is pruned by removing its branches. For example, Cost Complexity Algorithm.

46. What is meant by Pattern?

Ans. Pattern represents the knowledge.

47. Define the concept of prediction.

Ans. Prediction can be viewed as the construction and use of a model to assess the class of an unlabeled sample or to assess the value or value ranges of an attribute that a given sample is likely to have.

48. Define Clustering?

Ans. Clustering is a process of grouping the physical or conceptual data object into clusters.

49. What do you mean by Cluster Analysis?

Ans. A cluster analysis is the process of analyzing the various clusters to organize the different objects into meaningful and descriptive objects.

50. What are the fields in which clustering techniques are used?

Ans.

- Clustering is used in biology to develop new plants and animal taxonomies.
- Clustering is used in business to enable marketers to develop new distinct groups of their customers and characterize the customer group on basis of purchasing.
- Clustering is used in the identification of groups of automobiles insurance policy customer.

- Clustering is used in the identification of groups of house in a city on the basis of house type, their cost and geographical location.
- Clustering is used to classify the document on the web for information discovery.

51. What are the requirements of cluster analysis?

Ans. The basic requirements of cluster analysis are

- Dealing with different types of attributes.
- Dealing with noisy data.
- Constraints on clustering.
- Dealing with arbitrary shapes.
- High dimensionality
- Ordering of input data
- Interpretability and usability
- Determining input parameter and
- Scalability.

52. What are the different types of data used for cluster analysis?

Ans. The different types of data used for cluster analysis are interval scaled, binary, nominal, ordinal and ratio scaled data.

53. What are interval scaled variables?

Ans. Interval scaled variables are continuous measurements of linear scale. For example, height and weight, weather temperature or coordinates for any cluster. These measurements can be calculated using Euclidean distance or Minkowski distance.

54. Define Binary variables? And what are the two types of binary variables?

Ans. Binary variables are understood by two states 0 and 1, when state is 0, variable is absent and when state is 1, variable is present. There are two types of binary variables, symmetric and asymmetric binary variables. Symmetric variables are those variables that have same state values and weights. Asymmetric variables are those variables that have not same state values and weights.

55. Define nominal, ordinal and ratio scaled variables?

Ans. A nominal variable is a generalization of the binary variable. Nominal variable has more than two states, For example, a nominal variable, color consists of four states, red, green, yellow, or black. In Nominal variables the total number of states is N and it is denoted by letters, symbols or integers. An ordinal variable also has more than two states but all these states are ordered in a meaningful sequence. A ratio scaled variable makes positive measurements on a non-linear scale, such as exponential scale, using the formula

$$Ae^{Bt} \text{ or } Ae^{-Bt}$$

where A and B are constants.

56. What do u mean by partitioning method?

Ans. In partitioning method a partitioning algorithm arranges all the objects into various partitions, where the total number of partitions is less than the total number of objects. Here each partition represents a cluster. The two types of partitioning method are k-means and k-medoids.

57. Define CLARA and CLARANS?

Ans. Clustering in LARge Applications is called as CLARA. The efficiency of CLARA depends upon the size of the representative data set. CLARA does not work properly if any representative data set from the selected representative data sets does not find best k-medoids.

To recover this drawback a new algorithm, Clustering Large Applications based upon RANdomized search (CLARANS) is introduced. The CLARANS works like CLARA, the only difference between CLARA and CLARANS is the clustering process that is done after selecting the representative data sets.

58. What is Hierarchical method?

Ans. Hierarchical method groups all the objects into a tree of clusters that are arranged in a hierarchical order. This method works on bottom-up or top-down approaches.

59. Differentiate Agglomerative and Divisive Hierarchical Clustering?

Ans. Agglomerative Hierarchical clustering method works on the bottom-up approach. In Agglomerative hierarchical method, each object creates its own clusters. The single Clusters are merged to make larger clusters and the process of merging continues until all the singular clusters are merged into one big cluster that consists of all the objects. Divisive Hierarchical clustering method works on the top-down approach. In this method all the objects are arranged within a big singular cluster and the large cluster is continuously divided into smaller clusters until each cluster has a single object.

60. What is CURE?

Ans. Clustering Using Representatives is called as CURE. The clustering algorithms generally work on spherical and similar size clusters. CURE overcomes the problem of spherical and similar size cluster and is more robust with respect to outliers.

61. Define Chameleon method?

Ans. Chameleon is another hierarchical clustering method that uses dynamic modeling. Chameleon is introduced to recover the drawbacks of CURE method. In this method two clusters are merged, if the interconnectivity between two clusters is greater than the interconnectivity between the objects within a cluster.

62. Define Density based method?

Ans. Density based method deals with arbitrary shaped clusters. In density-based method, clusters are formed on the basis of the region where the density of the objects is high.

63. What is a DBSCAN?

Ans. Density Based Spatial Clustering of Application Noise is called as DBSCAN. DBSCAN is a density based clustering method that converts the high-density objects regions into clusters with arbitrary shapes and sizes. DBSCAN defines the cluster as a maximal set of density connected points.

64. What do you mean by Grid Based Method?

Ans. In this method objects are represented by the multi resolution grid data structure. All the objects are quantized into a finite number of cells and the collection of cells build the grid structure of objects. The clustering operations are performed on that grid structure. This method is widely used because its processing time is very fast and that is independent of number of objects.

65. What is a STING?

Ans. Statistical Information Grid is called as STING; it is a grid based multi resolution clustering method. In STING method, all the objects are contained into rectangular cells, these cells are kept into various levels of resolutions and these levels are arranged in a hierarchical structure.

66. Define Wave Cluster?

Ans. It is a grid based multi resolution clustering method. In this method all the objects are represented by a multidimensional grid structure and a wavelet transformation is applied for finding the dense region. Each grid cell contains the information of the group of objects that map into a cell. A wavelet transformation is a process of signaling that produces the signal of various frequency sub bands.

67. What is Model based method?

Ans. For optimizing a fit between a given data set and a mathematical model based methods are used. This method uses an assumption that the data are distributed by probability distributions. There are two basic approaches in this method that are

1. Statistical Approach
2. Neural Network Approach.

68. What is the use of Regression?

Ans. Regression can be used to solve the classification problems but it can also be used for applications such as forecasting. Regression can be performed using many different types of techniques; in actually regression takes a set of data and fits the data to a formula.

69. What are the reasons for not using the linear regression model to estimate the output data?

Ans. There are many reasons for that, One is that the data do not fit a linear model, It is possible however that the data generally do actually represent a linear model, but the linear model generated is poor because noise or outliers exist in the data. Noise is erroneous data and outliers are data values that are exceptions to the usual and expected data.

70. What are the two approaches used by regression to perform classification?

Ans. Regression can be used to perform classification using the following approaches

1. Division: The data are divided into regions based on class.

2. Prediction: Formulas are generated to predict the output class value.

71. What do u mean by logistic regression?

Ans. Instead of fitting a data into a straight line logistic regression uses a logistic curve. The formula for the univariate logistic curve is

$$P = \frac{e^{(C_0 + C_1 X_1)}}{1 + e^{(C_0 + C_1 X_1)}}$$

The logistic curve gives a value between 0 and 1 so it can be interpreted as the probability of class membership.

72. What is Time Series Analysis?

Ans. A time series is a set of attribute values over a period of time. Time Series Analysis may be viewed as finding patterns in the data and predicting future values.

73. What are the various detected patterns?

Ans. Detected patterns may include:

Trends : It may be viewed as systematic non-repetitive changes to the values over time.

Cycles : The observed behavior is cyclic.

Seasonal : The detected patterns may be based on time of year or month or day.

Outliers : To assist in pattern detection , techniques may be needed to remove or reduce the impact of outliers.

74. What is Smoothing?

Ans. Smoothing is an approach that is used to remove the nonsystematic behaviors found in time series. It usually takes the form of finding moving averages of attribute values. It is used to filter out noise and outliers.

75. Give the formula for Pearson's *r*.

Ans. One standard formula to measure correlation is the correlation coefficient *r*, sometimes called Pearson's *r*. Given two time series, X and Y with means X' and Y', each with n elements, the formula for *r* is

$$\frac{\Sigma(x_i - X')(y_i - Y')}{\left(\Sigma(x_i - X')^2\ \Sigma(y_i - Y')^2\right)^{1/2}}$$

76. What is Auto regression?

Ans. Auto regression is a method of predicting a future time series value by looking at previous values. Given a time series $X = (x_1, x_2, ..., x_n)$ a future value, x_{n+1}, can be found using

$$x_{n+1} = \xi + \varphi_n x_n + \varphi_{n-1} x_{n-1} + ... + \varepsilon_{n+1}$$

Here ε_{n+1} represents a random error, at time $n + 1$. In addition, each element in the time series can be viewed as a combination of a random error and a linear combination of previous values.

77. Define data warehouse?

Ans. A data warehouse is a repository of multiple heterogeneous data sources organized under a unified schema at a single site to facilitate management decision making.

(or)

A data warehouse is a subject-oriented, time-variant and nonvolatile collection of data in support of management's decision-making process.

78. What are operational databases?

Ans. Organizations maintain large database that are updated by daily transactions are called operational databases.

79. Define OLTP?

Ans. If an on-line operational database systems is used for efficient retrieval, efficient storage and management of large amounts of data, then the system is said to be on-line transaction processing.

80. Define OLAP?

Ans. Data warehouse systems serves users (or) knowledge workers in the role of data analysis and decision-making. Such systems can organize and present data in various formats. These systems are known as on-line analytical processing systems.

81. How a database design is represented in OLTP systems?

Ans. Entity-relation model

82. How a database design is represented in OLAP systems?

Ans. Star schema

Snowflake schema

Fact constellation schema

83. Write short notes on multidimensional data model?

Ans. Data warehouses and OLTP tools are based on a multidimensional data model. This model is used for the design of corporate data warehouses and department data marts. This model contains a Star schema, Snowflake schema and Fact constellation schemas. The core of the multidimensional model is the data cube.

84. Define data cube?

Ans. It consists of a large set of facts (or) measures and a number of dimensions.

85. What are facts?

Ans. Facts are numerical measures. Facts can also be considered as quantities by which we can analyze the relationship between dimensions.

86. What are dimensions?

Ans. Dimensions are the entities (or) perspectives with respect to an organization for keeping records and are hierarchical in nature.

87. Define dimension table?

Ans. A dimension table is used for describing the dimension. For example, a dimension table for item may contain the attributes item_ name, brand and type.

88. Define fact table?

Ans. Fact table contains the name of facts (or) measures as well as keys to each of the related dimensional tables.

89. What are lattice of cuboids?

Ans. In data warehousing research literature, a cube can also be called as cuboids. For different (or) set of dimensions, we can construct a lattice of cuboids, each showing the data at different level. The lattice of cuboids is also referred to as data cube.

90. What is apex cuboid?

Ans. The 0-D cuboid which holds the highest level of summarization is called the apex cuboid. The apex cuboid is typically denoted by all.

91. List out the components of star schema?

Ans.
- A large central table (fact table) containing the bulk of data with no redundancy.
- A set of smaller attendant tables (dimension tables), one for each dimension.

92. What is snowflake schema?

Ans. The snowflake schema is a variant of the star schema model, where some dimension tables are normalized thereby further splitting the tables in to additional tables.

93. List out the components of fact constellation schema?

Ans. This requires multiple fact tables to share dimension tables. This kind of schema can be viewed as a collection of stars and hence it is known as galaxy schema (or) fact constellation schema.

94. Point out the major difference between the star schema and the snowflake schema?

Ans. The dimension table of the snowflake schema model may be kept in normalized form to reduce redundancies. Such a table is easy to maintain and saves storage space.

95. Which is popular in the data warehouse design, star schema model (or) snowflake schema model?

Ans. Star schema model, because the snowflake structure can reduce the effectiveness and more joins will be needed to execute a query.

96. Define concept hierarchy?

Ans. A concept hierarchy defines a sequence of mappings from a set of low-level concepts to higher-level concepts.

97. Define total order?

Ans. If the attributes of a dimension which forms a concept hierarchy such as "street<city< province_or_state <country", then it is said to be total order.

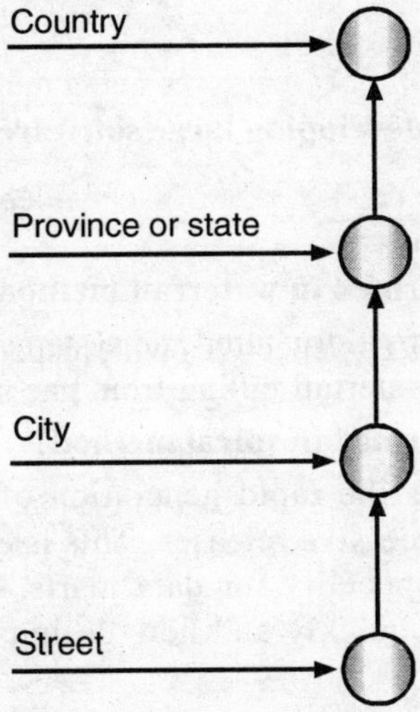

Fig. 2. Partial Order for Location.

98. Define partial order?

Ans. If the attributes of a dimension which forms a lattice such as "day<{month<quarter; week}<year, then it is said to be partial order.

99. Define schema hierarchy?

Ans. A concept hierarchy that is a total (or) partial order among attributes in a database schema is called a schema hierarchy.

100. List out the OLAP operations in multidimensional data model?

Ans.
- Roll-up
- Drill-down
- Slice and dice
- Pivot (or) rotate

101. What is roll-up operation?

Ans. The roll-up operation is also called drill-up operation which performs aggregation on a data cube either by climbing up a concept hierarchy for a dimension (or) by dimension reduction.

102. What is drill-down operation?

Ans. Drill-down is the reverse of roll-up operation. It navigates from less detailed data to more detailed data. Drill-down operation can be taken place by stepping down a concept hierarchy for a dimension.

103. What is slice operation?

Ans. The slice operation performs a selection on one dimension of the cube resulting in a sub cube.

104. What is dice operation?

Ans. The dice operation defines a sub cube by performing a selection on two (or) more dimensions.

105. What is pivot operation?

Ans. This is a visualization operation that rotates the data axes in an alternative presentation of the data.

106. List out the views in the design of a data warehouse?

Ans.
- Top-down view
- Data source view

- Data warehouse view
- Business query view

107. What are the methods for developing large software systems?

Ans.
- Waterfall method
- Spiral method

108. How the operation is performed in waterfall method?

Ans. The waterfall method performs a structured and systematic analysis at each step before proceeding to the next, which is like a waterfall falling from one step to the next.

109. How the operation is performed in spiral method?

Ans. The spiral method involves the rapid generation of increasingly functional systems, with short intervals between successive releases. This is considered as a good choice for the data warehouse development especially for data marts, because the turn around time is short, modifications can be done quickly and new designs and technologies can be adapted in a timely manner.

110. List out the steps of the data warehouse design process?

Ans.
- Choose a business process to model.
- Choose the grain of the business process
- Choose the dimensions that will apply to each fact table record.
- Choose the measures that will populate each fact table record.

111. Define ROLAP?

Ans. The ROLAP model is an extended relational DBMS that maps operations on multidimensional data to standard relational operations.

112. Define MOLAP?

Ans. The MOLAP model is a special purpose server that directly implements multidimensional data and operations.

113. Define HOLAP?

Ans. The hybrid OLAP approach combines ROLAP and MOLAP technology, benefiting from the greater scalability of ROLAP and the faster computation of MOLAP, *i.e.* a HOLAP server may allow large volumes of detail data to be stored in a relational database, while aggregations are kept in a separate MOLAP store.

114. What is enterprise warehouse?

Ans. An enterprise warehouse collects all the information's about subjects spanning the entire organization. It provides corporate-wide data integration, usually from one (or) more operational systems (or) external information providers. It contains detailed data as well as summarized data and can range in size from a few giga bytes to hundreds of giga bytes, tera bytes (or) beyond. An enterprise data warehouse may be implemented on traditional mainframes, UNIX super servers (or) parallel architecture platforms. It requires business modeling and may take years to design and build.

115. What is data mart?

Ans. Data mart is a database that contains a subset of data present in a data warehouse. Data marts are created to structure the data in a data warehouse according to issues such as hardware platforms and access control strategies. We can divide a data warehouse into data marts after the data warehouse has been created. Data marts are usually implemented on low-cost departmental servers that are UNIX (or) windows/NT based. The implementation cycle of the data mart is likely to be measured in weeks rather than months (or) years.

116. What are dependent and independent data marts?

Ans. Dependent data marts are sourced directly from enterprise data warehouses. Independent data marts are data captured from one (or) more operational systems (or) external information providers (or) data generated locally with in particular department (or) geographic area.

117. What is virtual warehouse?

Ans. A virtual warehouse is a set of views over operational databases. For efficient query processing, only some of the possible summary views may be materialized. A virtual warehouse is easy to build but requires excess capability on operational database servers.

118. Define indexing?

Ans. Indexing is a technique, which is used for efficient data retrieval (or) accessing data in a faster manner. When a table grows in volume, the indexes also increase in size requiring more storage.

119. What are the types of indexing?

Ans.
- B-Tree indexing
- Bit map indexing
- Join indexing

120. Define metadata?

Ans. Metadata is used in data warehouse is used for describing data about data *i.e.* meta data are the data that define warehouse objects. Metadata are created for the data names and definitions of the given warehouse.

121. Define VLDB?

Ans. **Very Large Data Base:** If a database whose size is greater than 100GB, then the database is said to be very large database.

122. What are the classifications of tools for data mining?

Ans.
- Commercial Tools
- Public domain Tools
- Research prototypes

123. What are commercial tools?

Ans. Commercial tools can be defined as the following products and usually are associated with the consulting activity by the same company:

1. 'Intelligent Miner' from IBM
2. 'SAS' System from SAS Institute
3. 'Thought' from Right Information Systems. etc

124. What are Public domain Tools?

Ans. Public domain Tools are largely freeware with just registration fees: 'Brute' from University of Washington. 'MC++' from Stanford university, Stanford, California.

125. What are Research prototypes?

Ans. Some of the research products may find their way into commercial market: 'DB Miner' from Simon Fraser University, British Columbia, 'Mining Kernel System' from University of Ulster, North Ireland.

126. What is the difference between generic single-task tools and generic multi-task tools?

Ans. Generic single-task tools generally use neural networks or decision trees. They cover only the data mining part and require extensive pre-processing and postprocessing steps. Generic multi-task tools offer modules for pre-processing and postprocessing steps and also offer a broad selection of several popular data mining algorithms as clustering.

127. What are the areas in which data warehouses are used in present and in future?

Ans. The potential subject areas in which data ware houses may be developed at present and also in future are:

1. Census data: The registrar general and census commissioner of India decennially compiles information of all individuals, villages, population groups, etc. This information is wide ranging such as the individual slip. A compilation of information of individual households, of which a database of 5%sample is maintained for analysis. A data warehouse can be built from this database upon which OLAP techniques can be applied, Data mining also can be performed for analysis and knowledge discovery

2. Prices of Essential Commodities: The ministry of food and civil supplies, Government of India complies daily data for about 300 observation centers in the entire country on the prices of essential commodities such as rice, edible oil etc, A data warehouse can be built for this data and OLAP techniques can be applied for its analysis

128. What are the other areas for Data warehousing and data mining?

Ans.
- Agriculture
- Rural development
- Health
- Planning
- Education
- Commerce and Trade.

129. Specify some of the sectors in which data warehousing and data mining are used?

Ans.
- Tourism
- Program Implementation
- Revenue
- Economic Affairs
- Audit and Accounts.

130. Describe the use of DBMiner.

Ans. Used to perform data mining functions, including characterization, association, classification, prediction and clustering.

131. Applications of DBMiner.

Ans. The DBMiner system can be used as a general-purpose online analytical mining system for both OLAP and data mining in relational database and datawarehouses. Used in medium to large relational databases with fast response time.

132. Give some data mining tools.

Ans. DBMiner
GeoMiner
Multimedia miner
WeblogMiner

133. Mention some of the application areas of data mining.

Ans. DNA analysis
Financial data analysis
Retail Industry
Telecommunication industry

Market analysis

Banking industry

Health care analysis.

134. Differentiate data query and knowledge query.

Ans. A data query finds concrete data stored in a database and corresponds to a basic retrieval statement in a database system. A knowledge query finds rules, patterns and other kinds of knowledge in a database and corresponds to querying database knowledge including deduction rules, integrity constraints, generalized rules, frequent patterns and other regularities.

135. Differentiate direct query answering and intelligent query answering.

Ans. Direct query answering means that a query answers by returning exactly what is being asked. Intelligent query answering consists of analyzing the intent of query and providing generalized, neighborhood, or associated information relevant to the query.

136. Define visual data mining.

Ans. Discovers implicit and useful knowledge from large data sets using data and/or knowledge visualization techniques. Integration of data visualization and data mining.

137. What does audio data mining mean?

Ans. Uses audio signals to indicate patterns of data or the features of data mining results. Patterns are transformed into sound and music. To identify interesting or unusual patterns by listening pitches, rhythms, tune and melody.

Steps involved in DNA analysis: Semantic integration of heterogeneous, distributed genome databases. Similarity search and comparison among DNA sequences. Association analysis: Identification of co-occuring gene sequences.

Path analysis: Linking genes to different stages of disease development. Visualization tools and genetic data analysis.

138. What are the factors involved while choosing data mining system?

Ans. Data types

System issues

Data sources

Data Mining functions and methodologies

Coupling data mining with database and/or data warehouse systems

Scalability

Visualization tools

Data mining query language and graphical user interface.

139. Define DMQL.

Ans. **Data Mining Query Language:** It specifies clauses and syntaxes for performing different types of data mining tasks for example data classification, data clustering and mining association rules. Also it uses SQl-like syntaxes to mine databases.

140. Define text mining.

Ans. Extraction of meaningful information from large amounts free format textual data. Useful in Artificial intelligence and pattern matching Also known as text mining, knowledge discovery from text, or content analysis.

141. What does web mining mean ?

Ans. Technique to process information available on web and search for useful data. To discover web pages, text documents, multimedia files, images, and other types of resources from web. Used in several fields such as E-commerce, information filtering, fraud detection and education and research.

142. Define spatial data mining.

Ans. Extracting undiscovered and implied spatial information.

Spatial data: Data that is associated with a location used in several fields such as geography, geology, medical imaging etc.

143. Explain multimedia data mining.

Ans. Mines large data bases. Does not retrieve any specific information from multimedia databases. Derive new relationships, trends, and patterns from stored multimedia data mining. Used in medical diagnosis, stock markets, Animation industry, Airline industry, Traffic management systems, Surveillance systems etc.

144. Explain the evolution of Database technology?

Ans.
- Data collection and Database creation
- Database management systems
- Advanced database systems
- Data warehousing and Data Mining
- Web-based Database systems
- New generation of Integrated information systems

145. Explain the steps of knowledge discovery in databases?

Ans.
- Data cleaning
- Data integration
- Data selection
- Data transformation
- Data mining
- Pattern evaluation
- Knowledge presentation

146. Explain the architecture of data mining system?

Ans.
- Database, datawarehouse, or other information repository
- Database or data warehouse server
- Knowledge base
- Data mining engine
- Pattern evaluation module
- Graphical user interface.

147. Explain various tasks in data mining?

(Or)

Explain the taxonomy of data mining tasks?

Ans.
- Predictive modeling
- Classification
- Regression
- Time series analysis
- Descriptive modeling
- Clustering
- Summarization

- Association rules
- Sequence discovery.

148. Explain various techniques in data mining?

Ans.
- Statistics (or) Statistical perspectives
- Point estimation
- Data summarization
- Bayesian techniques
- Hypothesis testing
- Correlation
- Regression
- Machine learning
- Decision trees
- Hidden markov models
- Artificial neural networks
- Genetic algorithms
- Meta learning

149. Explain the issues regarding classification and prediction?

Ans.
- Preparing the data for classification and prediction
  - Data cleaning
  - Relevance analysis
  - Data transformation
- Comparing classification methods
  - Predictive accuracy
  - Speed
  - Robustness
  - Scalability
  - Interpretability

150. Explain classification by Decision tree induction?

Ans.
- Decision tree induction
- Attribute selection measure.
- Tree pruning
- Extracting classification rules from decision trees

151. Write short notes on patterns?

Ans.
- Pattern definition
- Objective measures
- Subjective measures
- Can a data mining system generate all of the interesting patterns?
- Can a data mining system generate only interesting patterns?

152. Explain mining single –dimensional Boolean associated rules from transactional databases?

Ans.
- The apriori algorithm: Finding frequent itemsets using candidate generation
- Mining frequent item sets without candidate generation

153. Explain apriori algorithm?

Ans.
- Apriori property
- Join steps
- Prune step
- Example
- Algorithm

154. Explain how the efficiency of apriori is improved?

Ans.
- Hash-based technique (hashing item set counts)
- Transaction reduction (reducing the number of transactions scanned in future iteration)
- Partitioning (Partitioning the data to find candidate item sets)
- Sampling (mining on a subset of the given data)
- Dynamic item set counting (adding candidate item sets at different points during a scan)

155. Explain frequent item set without candidate without candidate generation?

Ans.
- Frequent patterns growth (or) FP-growth
- Frequent pattern tree (or) FP-tree
- Algorithm

156. Explain mining Multi-dimensional Boolean association rules from transaction databases?

Ans.
- Multi-dimensional (or) Multilevel association rules
- Approaches to mining Multilevel association rules
- Using uniform minimum support for all levels

Using reduced minimum support at lower levels

(*i*) Level-by-level independent

(*ii*) Level-cross filtering by single

(*iii*) Level-cross filtering by k-item set

Checking for redundant Multilevel association rules

157. Explain constraint-based association mining?

Ans.
- Knowledge type constraints
- Data constraints
- Dimension/level constraints
- Interestingness constraints
- Rule constraints
- Metarule-Guided mining of association of association rules
- Mining guided by additional rule constraints

158. Explain regression in predictive modeling?

Ans.
- Regression definition
- Linear regression
- Multiple regression
- Non-linear regression
- Other regression models

159. Explain statistical perspective in data mining?

Ans.
- Point estimation
- Data summarization

- Bayesian techniques
- Hypothesis testing
- Regression
- Correlation

160. Explain Bayesian classification.

Ans.
- Bayesian theorem
- Naïve Bayesian classification
- Bayesian belief networks
- Bayesian learning

161. Discuss the requirements of clustering in data mining.

Ans.
- Scalability
- Ability to deal with different types of attributes
- Discovery of clusters with arbitrary shape
- Minimal requirements for domain knowledge to determine input parameters
- Ability to deal with noisy data
- Insensitivity to the order of input records
- High dimensionality
- Interpretability and usability
- Interval scaled variables
- Binary variables
- Symmetric binary variables
- Asymmetric binary variables
- Nominal variables
- Ordinal variables
- Ratio-scaled variables

162. Explain the partitioning method of clustering.

Ans. K-means clustering
K-medoids clustering

163. Explain Visualization in data mining.

Ans. Various forms of visualizing the discovered patterns
- Rules
- Table
- Crosstab
- Pie chart
- Bar chart
- Decision tree
- Data cube
- Histogram
- Quantile plots
- q-q plots
- Scatter plots
- Loess curves

164. Discuss the components of data warehouse.

Ans.
- Subject-oriented
- Integrated
- Time-Variant
- Non-volatile

165. List out the differences between OLTP and OLAP.

Ans.
- Users and system orientation
- Data contents
- Database design
- View
- Access patterns

166. Discuss the various schematic representations in multidimensional model.

Ans.
- Star schema
- Snow flake schema
- Fact constellation schema

167. Explain the OLAP operations I multidimensional model.

Ans.
- Roll-up
- Drill-down
- Slice and dice
- Pivot or rotate

168. Explain the design and construction of a data warehouse.

Ans.
- Design of a data warehouse
- Top-down view
- Data source view
- Data warehouse view
- Business query view
- Process of data warehouse design

169. Explain the three-tier data warehouse architecture.

Ans.
- Warehouse database server(Bottom tier)
- OLAP server(middle tier)
- Client(top tier)

170. Explain indexing.

Ans.
- Definition
- B-Tree indexing
- Bit-map indexing
- Join indexing

171. Write notes on metadata repository.

Ans.
- Definition
- Structure of the data warehouse
- Operational metadata
- Algorithms used for summarization
- Mapping from operational environment to data warehouse

- Data related to system performance
- Business metadata

172. Write short notes on VLDB.

Ans.
- Definition
- Challenge related to database technologies
- Issues in VLDB

173. Explain data mining applications for Biomedical and DNA data analysis.

Ans.
- Semantic integration of heterogeneous, distributed genome databases
- Similarity search and comparison among DNA sequences
- Association analysis.
- Path analysis
- Visualization tools and genetic data analysis.

174. Explain data mining applications fro financial data analysis.

Ans.
- Loan payment prediction and customer credit policy analysis.
- Classification and clustering of customers fro targeted marketing.
- Detection of money laundering and other financial crimes.

175. Explain data mining applications for retail industry.

Ans.
- Multidimensional analysis of sales, customers, products, time and region.
- Analysis of the effectiveness of sales campaigns.
- Customer retention-analysis of customer loyalty.
- Purchase recommendation and cross-reference of items.

176. Explain data mining applications for Telecommunication industry.

Ans.
- Multidimensional analysis of telecommunication data.
- Fraudulent pattern analysis and the identification of unusual patterns.
- Multidimensional association and sequential pattern analysis
- Use of visualization tools in telecommunication data analysis.

177. Explain DBMiner tool in data mining.

Ans.
- System architecture
- Input and Output
- Data mining tasks supported by the system
- Support of task and method selection
- Support of the KDD process
- Main applications
- Current status

178. Explain how data mining is used in health care analysis.

Ans.
- Health care data mining and its aims
- Health care data mining technique
- Segmenting patients into groups
- Identifying patients into groups
- Identifying patients with recurring health problems
- Relation between disease and symptoms
- Curbing the treatment costs

- Predicting medical diagnosis
- Medical research
- Hospital administration
- Applications of data mining in health care
- Conclusion.

179. Explain how data mining is used in banking industry.

Ans.
- Data collected by data mining in banking
- Banking data mining tools
- Mining customer data of bank
- Mining for prediction and forecasting
- Mining for fraud detection
- Mining for cross selling bank services
- Mining for identifying customer preferences
- Applications of data mining in banking
- Conclusion

180. Explain the types of data mining.

Ans.
- Audio data mining
- Video data mining
- Image data mining
- Scientific and statistical data mining

181. What are the uses of statistics in data mining?

Ans. Statistics is used to
- to estimate the complexity of a data mining problem;
- suggest which data mining techniques are most likely to be successful; and
- identify data fields that contain the most "surface information".

182. What are the factors to be considered while selecting the sample in statistics?

Ans. The sample should be
- Large enough to be representative of the population.
- Small enough to be manageable.
- Accessible to the sampler.
- Free of bias.

183. Name some advanced database systems.

Ans. Object-oriented databases, Object-relational databases.

184. Name some specific application oriented databases.

Ans.
- Spatial databases,
- Time-series databases,
- Text databases and multimedia databases.

185. Define Relational databases.

Ans. A relational database is a collection of tables, each of which is assigned a unique name. Each table consists of a set of attributes (columns or fields) and usually stores a large set of tuples (records or rows). Each tuple in a relational table represents an object identified by a unique key and described by a set of attribute values.

186. Define Transactional Databases.

Ans. A transactional database consists of a file where each record represents a transaction. A Transaction typically includes a unique transaction identity number (trans_ID), and a list of the items making up the transaction.

187. Define Spatial Databases.

Ans. Spatial databases contain spatial-related information. Such databases include geographic (map) databases, VLSI chip design databases, and medical and satellite image databases. Spatial data may be represented in raster format, consisting of n-dimensional bit maps or pixel maps.

188. What is Temporal Database?

Ans. Temporal database store time related data .It usually stores relational data that include time related attributes. These attributes may involve several time stamps, each having different semantics.

189. What are Time-Series databases?

Ans. A Time-Series database stores sequences of values that change with time, such as data collected regarding the stock exchange.

190. Why machine learning is done?

Ans.
1. To understand and improve the efficiency of human learning.
2. To discover new things or structure that is unknown to human beings.
3. To fill in skeletal or computer specifications about a domain.

191. Give the components of a learning system.

Ans.
1. Critic
2. Sensors
3. Learning Element
4. Performance Element
5. Effectors
6. Problem generators.

192. What are the steps in the data mining process?

Ans.
1. Data cleaning
2. Data integration
3. Data selection
4. Data transformation
5. Data mining
6. Pattern evaluation
7. Knowledge representation

193. Define data cleaning.

Ans. Data cleaning means removing the inconsistent data or noise and collecting necessary information.

194. Define data mining.

Ans. Data mining is a process of extracting or mining knowledge from huge amount of data.

195. Define pattern evaluation.

Ans. Pattern evaluation is used to identify the truly interesting patterns representing knowledge based on some interesting measures.

196. Define knowledge representation

Ans. Knowledge representation techniques are used to present the mined knowledge to the user.

197. What is Visualization?

Ans. Visualization is for depiction of data and to gain intuition about data being observed. It Assists the analysts in selecting display formats, viewer perspectives and data representation schema.

198. Define Spatial Visualization.

Ans. Spatial visualization depicts actual members of the population in their feature space.

199. What is Descriptive and predictive data mining?

Ans. Descriptive data mining describes the data set in a concise and summertime manner and presents interesting general properties of the data. Predictive data mining analyzes the data in order to construct one or set of models and attempts to predict the behavior of new data sets.

200. What is Data Generalization?

Ans. It is process that abstracts a large set of task-relevant data in a database from a relatively low conceptual to higher conceptual levels two approaches for Generalization.

(*a*) Data cube approach

(*b*) Attribute-oriented induction approach.

201. Define Attribute Oriented Induction.

Ans. These method collets the task-relevant data using a relational database query and then perform generalization based on the examination in the relevant set of data.

202. What is bootstrap?

Ans. An interpretation of the jack knife is that the construction of pseudo value is based on repeatedly and systematically sampling with out replacement from the data at hand. This lead to generalized concept to repeated sampling with replacement called bootstrap.

203. View of statistical approach?

Ans. Statistical method is interested in interpreting the model. It may sacrifice some performance to be able to extract meaning from the model structure. If accuracy is acceptable then the reason that a model can be decomposed in to revealing parts is often more useful than a 'black box' system, especially during early stages of investigation and design cycle.

204. Define Deterministic models?

Ans. Deterministic models, which takes no account of random variables, but gives precise, fixed reproducible output.

205. Define Systems and Models?

Ans. System is a collection of interrelated objects and Model is a description of a system. Models are abstract, and conceptually simple.

206. How do you choose the best model?

Ans. All things being equal, the smallest model that explains the observations and fits the objectives that should be accepted. In reality, the smallest means the model should optimizes a certain scoring function (e.g. Least nodes, most robust, least assumptions).

207. What is clustering?

Ans. Clustering is the process of grouping the data into classes or clusters so that objects within a cluster have high similarity in comparison to one another, but are very dissimilar to objects in other clusters.

208. What are the requirements of clustering?

Ans.
- Scalability
- Ability to deal with different types of attributes
- Ability to deal with noisy data

- Minimal requirements for domain knowledge to determine input parameters
- Constraint based clustering
- Interpretability and usability.

209. State the categories of clustering methods?

Ans.
- Partitioning methods
- Hierarchical methods
- Density based methods
- Grid based methods
- Model based methods

210. What is linear regression?

Ans. In linear regression data are modeled using a straight line. Linear regression is the simplest form of regression. Bivariate linear regression models a random variable Y called response variable as a linear function of another random variable X, called a predictor variable.

$$Y = a + b X$$

211. State the types of linear model and state its use?

Ans. Generalized linear model represent the theoretical foundation on which linear regression can be applied to the modeling of categorical response variables. The types of generalized linear model are :

- Logistic regression
- Poisson regression

212. Write the preprocessing steps that may be applied to the data for classification and prediction.

Ans. (*a*) Data Cleaning
(*b*) Relevance Analysis
(*c*) Data Transformation.

213. Define Data Classification.

Ans. It is a two-step process. In the first step, a model is built describing a pre-determined set of data classes or concepts. The model is constructed by analyzing database tuples described by attributes. In the second step the model is used for classification.

214. What is a "decision tree"?

Ans. It is a flow-chart like tree structure, where each internal node denotes a test on an attribute, each branch represents an outcome of the test, and leaf nodes represent classes or class distributions. Decision tree is a predictive model. Each branch of the tree is a classification question and leaves of the tree are partition of the dataset with their classification.

215. Where are decision trees mainly used?

Ans. Used for exploration of dataset and business problems Data preprocessing for other predictive analysis Statisticians use decision trees for exploratory analysis.

216. What is Association rule?

Ans. Association rule finds interesting association or correlation relationships among a large set of data items, which is used for decision-making processes. Association rules analyzes buying patterns that are frequently associated or purchased together.

217. Define support.

Ans. Support is the ratio of the number of transactions that include all items in the antecedent and consequent parts of the rule to the total number of transactions. Support is an association rule interestingness measure.

218. Define Confidence.

Ans. Confidence is the ratio of the number of transactions that include all items in the consequent as well as antecedent to the number of transactions that include all items in antecedent. Confidence is an association rule interestingness measure.

219. How are association rules mined from large databases?

Ans. Association rule mining is a two-step process.

Find all frequent itemsets.

Generate strong association rules from the frequent itemsets.

220. What is the classification of association rules based on various criteria?

Ans. 1. Based on the types of values handled in the rule.

Boolean Association rule.

Quantitative Association rule.

2. Based on the dimensions of data involved in the rule.

(*a*) Single Dimensional Association rule.

(*b*) Multi Dimensional Association rule.

3. Based on the levels of abstractions involved in the rule.

Single level Association rule.

Multi level Association rule.

4. Based on various extensions to association mining.

Maxpatterns.

Frequent closed itemsets.

221. What are the advantages of Dimensional modeling?

Ans. 1. Ease of use.

2. High performance
3. Predictable, standard framework
4. Understandable
5. Extensible to accommodate unexpected new data elements and new design decisions.

222. Define Dimensional Modeling?

Ans. Dimensional modeling is a logical design technique that seeks to present the data in a standard framework that intuitive and allows for high-performance access. It is inherently dimensional and adheres to a discipline that uses the relational model with some important restrictions.

223. What comprises of a dimensional model?

Ans. Dimensional model is composed of one table with a multipart key called fact table and a set of smaller tables called dimension table. Each dimension table has a single part primary key that corresponds exactly to one of the components of multipart key in the fact table.

224. Define a data mart?

Ans. Data mart is a pragmatic collection of related facts, but does not have to be exhaustive or Exclusive. A data mart is both a kind of subject area and an application. Data mart is a collection of numeric facts.

225. What are the advantages of a data-modeling tool?

Ans.
- Integrates the data warehouse model with other corporate data models.
- Helps assure consistency in naming.
- Creates good documentation in a variety of useful formats.
- Provides a reasonably intuitive user interface for entering comments about objects.

226. What is data warehouse performance issue?

Ans. The performance of a data warehouse is largely a function of the quantity and type of data stored within a database and the query/data loading workload placed upon the system.

227. What are the types of performance issue?

Ans.
1. Capacity planning for the data warehouse
2. Data placement techniques within a data warehouse
3. Application Performance Techniques.
4. Monitoring the Data Warehouse.

228. Why do you need data warehouse life cycle process?

Ans. Data warehouse life cycle approach is essential because it ensures that the project pieces are brought together in the right order and at the right time.

229. What are the steps in the life cycle approach?

Ans.
- Project Planning
- Business Requirements definition
- **Data track:** Dimensional modeling, Physical Design, Data Staging Design and Development.
- **Technology track:** Technical Architecture design, Product Selection & Installation
- **Application track:** End user Application Specification, End user Application Development.
- Deployment
- Maintenance and Growth
- Project Management.

230. What are the merits of Data Warehouse ?

Ans.
- Ability to make effective decisions from database
- Better analysis of data and decision support
- Discover trends and correlations that benefits business
- Handle huge amount of data.

231. What are the characteristics of data warehouse?

Ans.
- Separate
- Available
- Integrated
- Subject Oriented
- Not Dynamic
- Consistency
- Iterative Development
- Aggregation Performance

232. List some of the Data Warehouse tools?

Ans.
- OLAP (Online Analytic Processing)
- ROLAP (Relational OLAP)
- End User Data Access tool
- Ad Hoc Query tool
- Data Transformation services
- Replication

233. Explain OLAP?

Ans. The general activity of querying and presenting text and number data from Data Warehouses, as well as a specifically dimensional style of querying and presenting that is exemplified by a number of "OLAP Vendors". The OLAP vendors technology is no relational and is almost always biased on an explicit multidimensional cube of data. LAP databases are also known as multidimensional cube of databases.

234. Explain ROLAP?

Ans. ROLAP is a set of user interfaces and applications that give a relational database a dimensional flavour. ROLAP stands for Relational Online Analytic Processing.

235. Explain End User Data Access tool?

Ans. End User Data Access tool is a client of the data warehouse. In a relational data warehouse, such a client maintains a session with the presentation server, sending a stream of separate SQL requests to the server. Evevtually the end user data access tool is done with the SQL session and turns around to present a screen of data or a report, a graph, or some other higher form of analysis to the user. An end user data access tool can be as simple as an Ad Hoc query tool or can be complex as a sophisticated data mining or modeling application.

236. Explain Ad Hoc query tool?

Ans. A specific kind of end user data access tool that invites the user to form their own queries by directly manipulating relational tables and their joins. Ad Hoc query tools, as powerful as they are, can only be effectively used and understood by about 10% of all the potential end users of a data warehouse.

237. Name some of the data mining applications?

Ans.
- Data mining for Biomedical and DNA data analysis
- Data mining for Financial data analysis
- Data mining for the Retail industry
- Data mining for the Telecommunication industry.

238. Name some of the data mining applications.

Ans.
- Data mining for Biomedical and DNA data analysis
- Data mining for Financial data analysis
- Data mining for the Retail industry
- Data mining for the Telecommunication industry.

239. What is the difference between "supervised" and unsupervised" learning scheme ?

Ans. In data mining during classification the class label of each training sample is provided, this type of training is called supervised learning *i.e.*, the learning of the model is supervised in that it is told to which class each training sample belongs. For example, Classification In unsupervised learning the class label of each training sample is not known and the member or set of classes to be learned may not be known in advance. For example, Clustering

240. Explain the various OLAP operations.

Ans. (*a*) **Roll-up:** The roll-up operation performs aggregation on a data cube, either by climbing up a concept hierarchy for a dimension.

(*b*) **Drill-down:** It is the reverse of roll-up. It navigates from less detailed data to more detailed data.

(*c*) **Slice:** Performs a selection on one dimension of the given cube, resulting in a sub cube.

241. Why is data quality so important in a data warehouse environment?

Ans. Data quality is important in a data warehouse environment to facilitate decision-making. In order to support decision-making, the stored data should provide information from a historical perspective and in a summarized manner.

242. How can data visualization help in decision-making?

Ans. Data visualization helps the analyst gain intuition about the data being observed. Visualization applications frequently assist the analyst in selecting display formats, viewer perspective and data representation schemas that faster deep intuitive understanding thus facilitating decision-making.

243. What do you mean by high performance data mining?

Ans. Data mining refers to extracting or mining knowledge. It involves an integration of techniques from multiple disciplines like database technology, statistics, machine learning, neural networks, etc. When it involves techniques from high performance computing it is referred as high performance data mining.

244. What is Data Mining? Explain the steps in Knowledge Discovery?

Ans. Data mining refers to extracting or mining knowledge from large amount of data. The steps in knowledge discovery are:

- Data cleaning
- Data integration
- Data selection
- Data transformation
- Data mining
- Pattern Evolution
- Knowledge Discovery.

245. Explain the data pre-processing techniques in detail?

Ans. The data preprocessing techniques are:

- Data Cleaning
- Data integration
- Data transformation
- Data reduction

246. Name any smoothing Techniques?

Ans.
- Binning
- Clustering
- Regression

247. Name Data transformation steps in detail?

Ans.
- Smoothing
- Aggregation
- Generalization
- Normalization
- Attribute Construction

248. Explain Normalization in detail?

Ans.
- Min Max Normalization
- Z-Score Normalization
- Normalization by decimal scaling

249. Name steps in data reduction?

Ans.
- Data cube Aggregation
- Attribute subset Selection
- Dimensional reduction
- Numerosity reduction.

250. Name parametric methods and non-parametric methods of reduction?

Ans. Parametric Methods:
- Regression Model
- Log linear Model

Non-Parametric Methods:
- Sampling
- Histogram
- Clustering

251. Explain Data Discrimination and Concept Hierarchy Generation?

Ans. Discrimination and concept hierarchy generation for numerical data:
- Segmentation by natural partitioning
- Binning
- Histogram Analysis
- Cluster Analysis

252. Explain Data mining Primitives?

Ans. There are five data mining primitives. They are:
- Task relevant data
- Kinds of knowledge to be mined
- Concept Hierarchies
- Interesting Measures
- Knowledge Presentation and Visualization Technique to be used for Discovery patterns.

253. Name the steps done while classification done by Decision tree induction?

Ans.
- Steps in decision tree induction
- Generation of decision tree algorithm
- Example and diagram
- Tree pruning

254. Explain the types of data in cluster analysis.

Ans.
- Data matrix
- Dissimilarity matrix
- Interval scaled variables
- Binary variables
- Nominal, Ordinal and Ratio scaled variables

255. Explain Outlier analysis?

Ans.
- Statistical based outlier detection
- Distance based outlier detection
- Deviation based outlier detection

256. Explain Mining complex types of data?

Ans.
- Multidimensional analysis and descriptive mining
- Mining spatial databases
- Mining Multimedia databases
- Mining Text databases
- Mining Time-series and sequence data
- Mining WWW

257. Briefly explain about Data Mining Application?

Ans.
- Financial Data Analysis
- Retail Industry
- Telecommunication Industry
- Biological Data Analysis
- Scientific Application

❑❑❑

TEST YOURSELF

SHORT ANSWER QUESTIONS

1. Define data warehouse?
2. What are operational databases?
3. Define OLTP?
4. Define OLAP?
5. How a database design is represented in OLTP systems?
6. How a database design is represented in OLAP systems?
7. Write short notes on multidimensional data model?
8. Define data cube?
9. What are facts?
10. What are dimensions?
11. Define dimension table?
12. Define fact table?
13. What are lattice of cuboids?
14. What is apex cuboid?
15. List out the components of star schema?
16. What is snowflake schema?
17. List out the components of fact constellation schema?
18. Point out the major difference between the star schema and the snowflake schema?
19. Which is popular in the data warehouse design, star schema model (or) snowflake schema model?
20. Define concept hierarchy?
21. Define total order?
22. Define partial order?
23. Define schema hierarchy?
24. List out the OLAP operations in multidimensional data model?
25. What is roll-up operation?
26. What is drill-down operation?
27. What is slice operation?
28. What is dice operation?
29. What is pivot operation?
30. List out the views in the design of a data warehouse?
31. What are the methods for developing large software systems?
32. How the operation is performed in waterfall method?
33. How the operation is performed in spiral method?
34. List out the steps of the data warehouse design process?
35. Define ROLAP?
36. Define MOLAP?
37. Define HOLAP?
38. What is enterprise warehouse?

39. What is data mart?
40. What are dependent and independent data marts?
41. What is virtual warehouse?
42. Define indexing?
43. What are the types of indexing?
44. Define metadata?
45. Define VLDB?
46. What are the classifications of tools for data mining?
47. What are the areas in which data warehouses are used in present and in future?
48. What are the other areas for Data warehousing and data mining?
49. Specify some of the sectors in which data warehousing and data mining are used?
50. Describe the use of DBMiner.
51. Applications of DBMiner.
52. Give some data mining tools.
53. Mention some of the application areas of data mining.
54. Differentiate data query and knowledge query.
55. Differentiate direct query answering and intelligent query answering.
56. Define visual data mining.
57. What does audio data mining mean?
58. What are the factors involved while choosing data mining system?
59. Define DMQL.
60. Define text mining.
61. What does web mining mean.
62. Define spatial data mining.
63. Explain multimedia data mining.
64. Define Data mining.
65. Give some alternative terms for data mining.
66. What is KDD.
67. What are the steps involved in KDD process.
68. What is the use of the knowledge base?
69. Mention some of the data mining techniques.
70. Give few statistical techniques.
71. What is meta learning.
72. Define Genetic algorithm.
73. What is the purpose of Data mining Technique?
74. Define Predictive model.
75. Data mining tasks that are belongs to predictive model.
76. Define descriptive model.
77. Define the term summarization.
78. Define cluster analysis.
79. Describe challenges to data mining regarding data mining methodology and user.
80. Define Association Rule Mining.
81. When we can say the association rules are interesting?
82. Explain Association rule in mathematical notations.

83. Define support and confidence in Association rule mining.
84. How are association rules mined from large databases?
85. Describe the different classifications of Association rule mining.
86. What is the purpose of Apriori Algorithm?
87. How to generate association rules from frequent item sets?
88. Give few techniques to improve the efficiency of Apriori algorithm.
89. What are the things suffering the performance of Apriori candidate generation technique.
90. Describe the method of generating frequent item sets without candidate generation.
91. Define Iceberg query.
92. Mention few approaches to mining Multilevel Association Rules
93. What are multidimensional association rules?
94. Define constraint-Based Association Mining.
95. Define the concept of classification.
96. What is Decision tree?
97. What is Attribute Selection Measure?
98. Describe Tree pruning methods.
99. Define Pre Pruning
100. Define Post Pruning.
101. What is meant by Pattern?
102. Define the concept of prediction.
103. Describe challenges to data mining regarding performance issues.
104. Describe issues relating to the diversity of database types.
105. How is a data warehouse different from a database?
106. What do you mean by Cluster Analysis?
107. What are the fields in which clustering techniques are used?
108. What are the requirements of cluster analysis?
109. What are the different types of data used for cluster analysis?
110. What are interval scaled variables?
111. What is Hierarchical method?
112. Differentiate Agglomerative and Divisive Hierarchical Clustering
113. Define Density based method?
114. What is a DBSCAN?
115. What do you mean by Grid Based Method?
116. What is Model based method?
117. What is the use of Regression?
118. What is Time Series Analysis?

10 OR 20 MARKS QUESTIONS

1. Explain data mining architecture
2. Explain DBMiner tool in data mining.
3. Explain how data mining is used in health care analysis.
4. Explain how data mining is used in banking industry.
5. Explain the types of data mining.

6. Explain the evolution of Database technology?
7. Explain the steps of knowledge discovery in databases?
8. Explain the architecture of data mining system?
9. Explain various tasks in data mining?
10. Explain the taxonomy of data mining tasks?
11. Explain various techniques in data mining.
12. Explain the issues regarding classification and prediction?
13. Explain classification by Decision tree induction?
14. Write short notes on patterns?
15. Explain mining single-dimensional Boolean associated rules from transactional databases?
16. Explain apriori algorithm?
17. Explain how the efficiency of apriori is improved?
18. Explain frequent item set without candidate without candidate generation?
19. Explain mining Multi-dimensional Boolean association rules from transaction
20. Explain constraint-based association mining?
21. Define Clustering?
22. Explain regression in predictive modeling?
23. Explain statistical perspective in data mining?
24. Explain Bayesian classification.
25. Discuss the requirements of clustering in data mining.
26. Explain the partitioning method of clustering.
27. Explain Visualization in data mining
28. Discuss the components of data warehouse.
29. List out the differences between OLTP and OLAP.
30. Discuss the various schematic representations in multidimensional model.
31. Explain the OLAP operations I multidimensional model.
32. Explain the design and construction of a data warehouse.
33. Expalin the three-tier data warehouse architecture.
34. Explain indexing.
35. Write notes on metadata repository.
36. Write short notes on Virtual DW.

❑❑❑

MODEL TEST PAPERS

MODEL TEST PAPER–1

B.TECH.
DATA MINING AND DATA WAREHOUSING

Time : 3 hours *Maximum Marks : 100*

Question 1–10 is compulsory, explain any four from 11-15.

1. Write down the applications of data warehousing. **(2 × 10 = 20)**
2. When is data mart appropriate?
3. What is concept hierarchy? give an example.
4. What do you mean by web mining.
5. Write the two measures of Association Rule.
6. Define conditional pattern base.
7. List out the major strength of decision tree method.
8. Distinguish between classification and clustering.
9. Define a spatial database.
10. List out any two various commercial data mining tools.
11. (*a*) (*i*) With a neat sketch explain the architecture of a data warehouse. **(20)**
(*ii*) Discuss the typical OLAP operations with an example.

Or

(*b*) (*i*) Discuss how computations can be performed efficiently on data cubes.
(*ii*) Write short notes on data warehouse meta data.
12. (*a*) (*i*) Explain how neural network approach is used in classification in detail. **(20)**
(*ii*) Give an account on data mining Query language.

Or

(*b*) Discuss text mining?
13. (*a*) Write and explain the algorithm for mining frequent item sets without candidate generation. Give relevant example. **(20)**

Or

(*b*) Discuss the approaches for mining multi level association rules from the transactional databases. Give relevant example.
14. (*a*) (*i*) Explain the algorithm for constructing a decision tree from training samples. **(20)**
(*ii*) Explain Bayes theorem.

Or

(*b*) Explain any two clustering methods in detail.
15. (*a*) (*i*) What is a multimedia database? Explain the methods of mining multimedia database? **(20)**

Or

(*b*) (*i*) Discuss the social impacts of data mining.
(*ii*) Discuss spatial data mining.

MODEL TEST PAPER–2

B.TECH.
DATA MINING AND DATA WAREHOUSING

Time : 3 hours *Maximum Marks : 100*

Note: Attempt ***all*** the questions.

1. Attempt any **four** parts of the questions :

(*a*) What is data warehouse ? What are the goals of data warehouse ?

(*b*) What do you mean by Granularity ? What is partitioning ?

(*c*) Compare and contrast OLTP and data warehouse.

(*d*) How are data marts different from data warehouse ? What are the different types of data marts ?

(*e*) What does web mining mean.

(*f*) Define spatial data mining.

2. Attempt any **two** of the following :

(*a*) What are the different components of a data warehouse ? Explain with the help of a diagram.

(*b*) What is metadata and why is it important ? Discuss the multidimensional data.

(*c*) Explain the steps of knowledge discovery in databases ?

3. Attempt any **two** of the following :

(*a*) What is on-line transaction processing (OLTP) ? Describe the evolution of OLTP. What are the critical features of OLTP systems ?

(*b*) Describe different Reporting and Query Tools and Applications.

(*c*) How Artificial Intelligence is used in Data warehousing. Explain with suitable examples.

4. Attempt any **four** of the following :

(*a*) What is data mining ? Describe the evolution of data mining.

(*b*) Explain apriori algorithm?

(*c*) Explain Bayesian classification.

(*d*) Explain the advantages of data mining.

(*e*) What are technologies used in data mining ?

(*f*) What do you mean by decision trees ? Explain.

5. Attempt any **two** of the following :

(*a*) Give E.F. Codd's 12 Guidelines for OLAP.

(*b*) What are genetic algorithms ? Explain taking one algorithm.

(*c*) Discuss data visualization with reference to Data Mining.

MODEL TEST PAPER–3

B.TECH.
DATA MINING AND DATA WAREHOUSING

Time : 3 hours *Maximum Marks : 100*

Note: Attempt *all* the questions.

1. Attempt any **two** of the following :

(*a*) (*i*) Define the data warehouse.

(*ii*) How is a data warehouse different from a database ? How are they similar ?

(*b*) (*i*) What are the different phases of the knowledge discovery from data base ?

(*ii*) What data mining attepts to facilitate ?

(*c*) What are schemas. Discuss various schemas used in data warehouse.

2. Attempt any **two** of the following :

(*a*) What is OLAP ? Explain various OLAP operations used on a data cube.

(*b*) Apply the Apriori algorithm on the following data set and find the maximum frequent item set. Set having support **20%** or more are frequent item set.

Show two association rules that have a confidence 70% or greater.

Trans ID	Items Purchased Sold
10	I1, I2, I4
20	I1, I7
30	I7, I5
40	I1, I2, I5
50	I6, I4
60	I6
70	I6, I7
80	I1, I2, I3, I4
90	I3, I5
00	I1, I2

(*c*) Discuss Association rules with their classification.

3. Attempt any **four** of the following :

(*a*) What are classification rules and how are decision trees related to them ?

(*b*) What is data classification ? How it is differ than prediction ?

(*c*) Describe the **ID3** algorithm of the decision tree construction. Why is it unsuitable for data mining applications ?

(*d*) Explain the issues regarding classification and prediction?

(*e*) Describe neural networks techniques for data mining. What are the main difficulties in using these techniques ?

(*f*) What is Bayesian classification ? How it classifies the Input data ?

4. Attempt any **two** of the following :

(*a*) How does clustering differ from classification ?

(*b*) What is supervised and unsupervised learning ? Why clustering is known as unsupervised learning ?

(*c*) Describe the genetic algorithms as data mining techniques. What are the main difficulties in using these techniques ?

5. Attempt any **two** of the following :

(*a*) What is backpropagation Neural Network topology ? How it is used in classification ?

(*b*) Compare hierarchical clustering and non hierarchical clustering algorithm. Explain the advantages and disadvantages over each other.

(*c*) Write differences between the Nearest Neighbour Data Mining Techniques and clustering.

MODEL TEST PAPER–4

B.TECH.
DATA MINING AND DATA WAREHOUSING

Time : 3 hours *Maximum Marks : 100*

Note: Attempt ***all*** the questions. Each questions carry equal marks.

1. Attempt any **two** of the following :

(*a*) Discuss the components of data warehouse.

(*b*) List out the differences between OLTP and OLAP.

2. Explain Testing and Tuning of a data warehouse.

3. (*a*) Expalin the three-tier data warehouse architecture.

(*b*) Write notes on metadata repository.

4. (*a*) Explain data mining architecture.

(*b*) Explain the issues regarding classification and prediction?

5. (*a*) Explain apriori algorithm? And how the efficiency of apriori is improved?

(*b*) Explain the partitioning method of clustering.

6. (*a*) Explain mining Multi-dimensional Boolean association rules from transaction.

(*b*) How association rules can be classified.

7. Discuss how ANN is useful in classification and what are the issues using this approach.

8. Attempt any **four** of the following :

(*a*) Define Clustering?

(*b*) Explain Bayesian classification.

(*c*) Virtual DW

(*d*) DMQL

(*e*) Spatial data, spatial DW

❑❑❑

EXAMINATION PAPERS

B.E. SEVENTH SEMESTER (IT-401-E)
EXAMINATION DECEMBER 2011
DATA WAREHOUSING AND DATA MINING

Time : 3 hours *Maximum Marks : 100*

Before answering the questions, candidates should ensure that they have been supplied the correct and complete question paper. No complain in this regard, will be entertained after examination.

Note: Attempt any ***five*** questions.

1. (*a*) Differentiate between DBMS and data warehouse. Also explain the concept of data cubes in detail. **10**

(*b*) Compare and contrast the star and snowflakes model. **10**

2. Explain the three tier architeture of data warehouse in detail. Also discuss the concept of virtual data warehouse. **20**

3. (*a*) Explain the various design issues during the implementation of data warehouse. **10**

(*b*) How tuning and testing of data warehouse has been done ? **10**

4. (*a*) Explain the various data mining techniques in detail. **10**

(*b*) Define the following terms : **10**

KDD and Data mining tools.

5. (*a*) How we specify the data during the query of data mining. Explain with a suitable example. **10**

(*b*) Explain the concept of pattern presentation and visualization specification in data mining ? **10**

6. (*a*) Explain the use of neural network for data mining in brief. **10**

(*b*) Explain the support vector machines in detail. **10**

7. (*a*) Compare and contrast the mining techniques for spatial and multimedia databases in detail. **10**

(*b*) What are the various issues in mining the WWW content ? **10**

8. Write short notes on the following :

(*a*) OLAP **5**

(*b*) Meta data **5**

(*c*) Data warehouse back end tools **5**

(*d*) Complex aggregation. **5**

B.E. SEVENTH SEMESTER (IT-401-E)
EXAMINATION MAY 2012
DATA WAREHOUSING AND DATA MINING

Time : 3 hours *Maximum Marks : 100*

Before answering the questions, candidates should ensure that they have been supplied the correct and complete question paper. No complain in this regard, will be entertained after examination.

Note: Attempt any *five* questions.

1. (*a*) Differentiate between Data mart and data warehouse. Also explain the various schemes for multidimensional databases in detail. **10**

(*b*) Compare and contrast the star and snowflakes model. **10**

2. Explain the three tier architeture of data warehouse in detail. Also discuss the concept of distributed data warehouse. **20**

3. (*a*) Differentiate between the following :

(*i*) OLTP Vs OLAP

(*ii*) ROLAP Vs MOLAP. **10**

(*b*) How tuning and testing of data warehouse has been done ? **10**

4. (*a*) Explain the various steps of knowledge discovery in detail. **10**

(*b*) Explain the various data mining techniques in detail. **10**

5. (*a*) What is the role of data specification and data hierarchy specification in data mining ? **10**

(*b*) Explain the concept of pattern presentation and visualization specification in data mining ? **10**

6. (*a*) Explain the use of genetic algorithm for data mining in brief. **10**

(*b*) Explain how fuzzy techniques are important in data mining. **10**

7. (*a*) Comapre and contrast the mining techniques for spatial and multimedia databases in detail. **10**

(*b*) What are the various issues in mining the Text Databases ? **10**

8. Write short notes on the following :

(*a*) Virtual Data warehouse. **5**

(*b*) Meta data **5**

(*c*) Neural Network **5**

(*d*) Data Cubes. **5**

❑❑❑

INDEX

A

A Data Mining Query Language, DMQL: Language Primitives 43
A Rough Set Based KDD Process 146
Ad Hoc Queries 76
Advantages and Disadvantages of Data Granularity 21
Advantages and Disadvantages of Data Mining 96
Advantages and Disadvantages of Data warehouse 15
Advantages and Disadvantages of Decision Trees 137
Advantages and Disadvantages of Fuzzy Logic 148
Advantages of Data Mining 96
Advantages of MOLAP 58
Advantages of ROLAP 59
Advantages of Virtual Warehouse 30
Aggregate Tables 44
Applications of Clustering 175
Applications of Data Mining 97
Apriori Algorithm 161
Architecture of Data Warehouse 26
Architecture of DM Systems 109
Association Rules 157
Automated Discovery of Previously Unknown Patterns 90
Automated Prediction of Trends and Behaviours 90

B

Background Knowledge 105
Basic Approaches for Mining Multidimensional Association Rules 169
Basic Data Mining Tasks 84
Bayes Theorem 129
Bayes' Theorem Derived via Conditional Probabilities 132
Bayesian Classification 129
Bayesian Network 135
Benefits of a Data Warehouse 6
Bottom Tier or Data Warehouse Server 28
Business Users Test Business Intelligence 74

C

C4.5 Algorithm 140
Can the Data Warehouse Perform? 74
CART: Classification and Regression Trees 142
Case Based Reasoning (CBR) 129
Categories of clustering methods 177
Certainty Factor 135
Challenges of Data Warehouse Testing: 73
Characteristics of a Data Warehouse 4
Characteristics of Metadata Architecture 24
Characteristics of Spatial Data Mining 189
Classical Portioning Method-k-means Method 177
Classification and Issues 115
Classification of Association Rules 160
Classification 115
Clustering 174
Comments on PAM 180
Complex Aggregation at Multiple Granularities: Multifeature Cube 72
Complex Data Objects Mining 209
Components, Architecture, Design and Construction of Data Warehouse 9
Concept of Frequent Itemsets, Closed Itemsets and Association Rules 158
Constraint Based Association Mining 170
Construction and Mining of Object Cubes 210
Contrasting OLTP and Data Warehousing Environments 7
Conversion from ER Model to Data Warehouse Architecture (Star Schema) 50
Cost Complexity Pruning 138
Cube Computation 67

D

Data Cubes 45
Data Granularity 20
Data Marts 30
Data Mining Application 88
Data Mining Query Language 103

Data Mining Techniques 91
Data Mining Tools Comparison 92
Data Mining Vs. Text Mining 197
Data Staging 'Area' 9
Data Warehouse 2
Data Warehouse Architecture 10
Data Warehouse Definitions 3
Data Warehouse Implementation 64
Data Warehouse Models 29
Data Warehouse Schema Architecture [MDU Dec. 2011; May 2012] 39
Data Warehouse vs. OLTP Transaction Systems 8
Data Warehouse: Features and Architecture 18
Data Warehousing Test Plan 74
Data Warehousing Testing Responsibilities 73
Data Warehousing Tools 71
Database Attributes can be Categorical or Quantitative. 169
Databases may be Huge 205
DBMS Vs. Data warehouse 6
Decision Tree Based Algorithm (ID3, C4.5, CART) 136
Decision Tree Based Algorithms 139
Defining a Snowflake Schema in DMQL 43
Defining a Star Schema in DMQL 43
Defining Data Mining 83
Density Based Clustering Methods 182
Dependent and Independent Data Marts 31
Design of Data Warehouse 26
Difference between KDD Vs. Data Mining 85
Difference between ROLAP AND MOLAP based techniques 68
Difference between ROLAP and MOLAP 59
Difference Between Star Schema and Snowflake Schema 41
Differences Between OLTP and OLAP System 55
Different Approaches For Mining Multilevel Association Rule 168
Different Text Mining Approaches 199
Difficult Training Set 204
Dimension Hierarchies 65
Dimensional Model vs ER Model 38
Dimensional Modeling 37
Disadvantages of Data Mining 97
Disadvantages of MOLAP 58
Disadvantages of ROLAP 59
Distance based Algorithms 128
DMQL: Syntax For Specification of Task-Relevant Data 106
DOLAP—Desktop OLAP 61
Dynamic Nature of Database 205

E

Efficient Data Cube Computation 64
End User Tools and Applications 10
Enterprise Warehouse 29
ER vs DM 50
ETL Process 2
ETL-Data Extraction 10
ETL-Data Transformation 10
ETL-Presentation/Loaded 'Area' 10
Example database with 4 items and 5 transactions 159

F

Fact Constellation Schema/Glaxy Schema 42
Features of Data Warehouse 18
Features of Multifeature Cube 72
Fixed Queries 76
Four major types of concept hierarchy are: 105
FP-Growth Algorithm 166
Future of Data Warehousing 79
Fuzzy Sets Approach 147

G

Generalization of Class Composition Hierarchies 210
Generalization of Object Identifiers and Class/ Subclass Hierarchies 210
Generalization-Based Mining of Plan Databases by Divide-and-Conquer 210
Generating Rules From Frequent Itemsets 162
Genetic Algorithm 142

H

Hierarchical Methods 180
History of Data Mining 82
History of Data Warehousing 2
HOLAP—Hybrid OLAP 60
How Data Mining Works 92

I

IBM–Intelligent Miner 93
Iceberg Cube 68
Iceberg Queries 167
ID3: ID3 stands for Iterative Dichotomiser 3 139

Importance of Data Mining Query Language 104
Indexing OLAP Data 69
Information Catalog 78
Introduction to Data Mining 81
Introduction to Data Warehousing 1
Issues and Challenges in Data Mining 204
Issues in Decision Trees Algorithms 138

K

KDD Vs. Data Mining 85
KDD-Data Mining Issues and Challenges 99
k-medoids algorithm 180
K-Medoids Method 179

L

Lattice of Cuboids 64

M

Materialization 66
m-Dimensional Array 46
Meta Data 10
Metadata Lifecycle 25
Metadata Repository 25
Metadata 22
Middle Tier or OLAP Server 28
Mining Complex Types of Data 188
Mining Multi Dimensional Association Rules from Relational Databases and Data Warehouse 169
Mining time series and sequence data 204
Mining Time Series Data 206
Model Based Clustering Methods 185
MOLAP Base Ccubing Algorithm 67
MOLAP—Multidimensional OLAP 57
Multidimensional Analysis of Multimedia Data 202
Multimedia Databases Mining 201
Multimedia Indexing and Retrieval System 202
Multimedia MINER 203

N

Naive Bayes Classifier 130
Need for a DM 38
Need of Data Warehouse 5
Need of Meta Data 23
NeoVista Software–Decision Series 96
Neural Networks 118
Noisy and Missing Data 204
OLAP and OLTP 54
OLAP Applications 61
OLAP operations in Multidimensional Data Model 56
OLAP Server/Models 57
OLAP Strategies for Large Data Warehouse Projects 52
OLTP vs. OLAP 54
Operations on Data Cubes 47
OPTICS: Ordering Points to Identify the Clustering Structure 184
Outlier Analysis 185

P

Partitioning Algorithms 177
Processing Queues 77
Properties of SVM 153
Pseudocode 141

Q

Query Management Facility 79
Query Manager 77
Query Patroller 78
Query Processing 47
Query Tuning 76

R

Reasons for Creating a Data Mart 31
Representation of Data Cubes 46
Representation of Totals 46
Requirements of Clustering in Data Mining 176
ROLAP-Based Cubing Algorithms 67
ROLAP—relational OLAP 58
Rough Sets Application 147
Rough Sets 145
Rule Strength Measures 158

S

SAS Institute–Enterprise Miner 95
Sequence Pattern Mining 211
Silicon Graphics–MineSet 95
Similarities and Dissimilarities Between DW and Data Marts 31
Simple Statement of Theorem 130
Snowflake Schema 40
Source Systems and Databases 9
Spatial Classification and Spatial Trend Analysis 191
Spatial Clustering Methods 191
Spatial database mining 188

SPSS–Clementine 94
Star Join Processing 48
Star Schema 39
Statistical Based Algorithm (Bayesian Classification) 129
Strengths of Dimensional Model 49
Summary of Apriori Algorithm 163
Support Victor Machine 148
SVM-General Concept 149
SVM—When Data Is Linearly Separable 151
Syntax for Interestingness Measure Specification 108
Syntax for Pattern Presentation and Visualization Specification 109
Syntax for Specifying the Kind of Knowledge to be Mined 107
Syntax for Task-Relevant Data Specification 104
Syntax for Task-Relevant Data Specifications 104

T

Techniques in Spatial DM 190
Testing and Tuning the Data Warehouse 73
Text Mining 196
Text Mining Application 200
The Advantages of Data Warehouse 26
The Data Warehouse Center 78
The Naive Bayes Probabilistic Model 130
The Need For Multidimensional Information 53
The Scope of Data Mining 90
The Steps in Implementing a Data Mart 33
Three Tier Architecture 27
Top Tier or Font End Tool Analysis 29
Training Bayesian Networks: Several Scenarios 136
Trends in Data Mining 206
Tuning Strategies 75
Tuning The Data Warehouse 75
Two Aspects of Time Series Analysis 209
Two Tier Data Warehouse Architecture—Independent Data Marts 27
Types of Knowledge to be Mined 105
Typical Kernel Functions 153

U

Uses of Spatial DM 190
Uses of Web Mining 193

V

Virtual Warehouse 30
Visualization of Discovered Patterns 106

W

Web Content Mining 193
Web Mining 192
Web Structure Mining 195
Web Usage Mining 194
What can a Data Warehouse do? 3
What can a Data Warehouse not do? 4
What is Metadata? 22
Why Use Data Mining? 92

❑❑❑

Win Prizes

ATTENTION READERS

Win Prizes

We request for your frank assessment regarding various aspects of the book as given below :

Data Mining and Data Warehousing

By : GUNJAN GOSWAMI

(MDU)

Please Fill up the Sheet in Neat and Clean Writing

(i) What topics of your syllabus, important for your examination point of view, are not covered in this book ?

....

....

....

....

(ii) What are the chapters wherein the treatment of the subject matter is not systematic or organised or updated ?

....

....

....

....

(iii) Have you come across misprints/mistakes/factual inaccuracies in the book ? Please specify the chapters and the page numbers.

....

....

....

....

(iv) What other books on the same subject have you found/heard better than the present book ? Please specify in terms of price and quality.

....

....

....

....

(v) Further suggestions and comments for the improvement of this book.

....

....

....

....

(PS : If need be please attach a separate sheet to write your views)

OTHER DETAILS

(i) Who recommended you this book ? (Please tick (✓) in the box near the option relevant to you.

☐ Teacher ☐ Friends ☐ Bookseller

....

....

....

....

(ii) Name of the recommending teacher, his designation and address:

....

....

....

....

(iii) Name and address of the bookseller you purchased the book from :

....

....

....

....

(iv) Name and address of your institution (Please mention the University or Board as the case may be)

....

....

....

....

(v) Your name and complete address :

....

....

....

....

(vi) Course you are enrolled in :

....

....

....

....

The best assessment will be awarded each month : The award will be in the form of our publications as decided by the Editorial Board.

Please mail the filled up coupon at the vearliest to :

S.K. KATARIA & SONS®
4885/109, Prakash Mahal,
Dr. Subhash Bhargav Lane,
Opposite Delhi Medical Association,
Daryaganj, New Delhi-110002
Phone: +91-11-23243489, 23269324
TeleFax : +91-11-23243489
email : katariabooks@yahoo.com